ADHESIVES IN MODERN MANUFACTURING

ADHESIVES

AUTHORS

In organizing, developing, and writing this book, the co-authors listed below have generously donated their time, effort, and professional skills to contribute to the advancement of manufacturing technology.

CHAIRMAN

Douglas F. Weyher, Supervisor,
Adhesives-Sealants,
Chemical Division,
Chrysler Corporation,
Trenton, Michigan

Jack Borsellino, Industrial Business Manager,
Chemical Division,
Thiokol Chemical Corporation,
Trenton, New Jersey

J. A. Clarke, Research Specialist,
The Dow Chemical Company,
Freeport, Texas

Leon F. Fiedler, Field Engineer,
B. F. Goodrich Company,
Akron, Ohio

Dr. B. R. Garrett, Product Manager,
Hexcel Corporation,
Dublin, California

Published by
Society of Manufacturing Engineers
Dearborn, Michigan, 1970

n modern manufacturing

Dr. Warren F. Hale, Project Scientist,
Research and Development Department,
Union Carbide Corporation,
Bound Brook, New Jersey

Daniel J. Horvath, Jr., Principal Design Engineer,
Body Engineering Office,
Ford Motor Company,
Dearborn, Michigan

Rudolph B. Janota, Director of Customer
and Technical Services,
Adhesive and Coating Department,
Swift Chemical Company,
Division of Swift & Company,
Oakbrook, Illinois

Leroy E. Meade, Aircraft Development
Engineer Specialist-Composite Materials,
Lockheed-Georgia Company,
Division of Lockheed Aircraft Corporation,
Marietta, Georgia

Julian R. Panek, Manager of
Technical Communications,
Chemical Division,
Thiokol Chemical Corporation,
Trenton, New Jersey

Dr. Cumming Paton, Market Manager,
Organic Division,
Monsanto Company,
St. Louis, Missouri

Harry E. Pattee, Senior Engineer,
Battelle Memorial Institute,
Columbus, Ohio

Carl B. Penn, President,
Pyles Industries, Inc.,
Subsidiary of Kent-Moore Corporation,
Wixom, Michigan

Max B. Rittgers, Supervisor,
Material Applications,
Gray Company, Incorporated,
Minneapolis, Minnesota

C. W. Roop, Manager, Applied Research,
Chemical Division,
USM Corporation,
Middleton, Massachusetts

Leonard R. Vertnik, Section Leader,
Technical Service-Adhesive Section,
Chemical Division,
General Mills, Incorporated,
Minneapolis, Minnesota

E. J. Bruno, Editor

Adhesives in Modern Manufacturing

Library of Congress Catalog Card Number: 79-93212
SBN 87263-017-X

FIRST EDITION ■ MANUFACTURING DATA SERIES
MANUFACTURED IN THE UNITED STATES OF AMERICA

PREFACE

The Society of Manufacturing Engineers, recognizing a need for a single source of information on adhesive joining for manufacturing engineers, formed a committee composed of men long experienced in this field to write a book discussing typical problems, applications, and equipment used to apply the adhesives. With this purpose in mind, the co-authors have written *Adhesives in Modern Manufacturing* to provide basic and applied information on adhesives and those related processes and equipment required for adhesive applications.

The book divides the field of adhesives into six basic types: chemical reactive types, evaporation or diffusion types, hot melt types (cooling), delayed-tack types, film types (heat and pressure), and pressure sensitive types. It discusses design considerations for the selection and use of these six adhesive types and their various modifications. Surface preparation for bonding, methods of bonding, and inspection and quality control are discussed. A number of case histories are also presented to illustrate the development and solution of practical problems in manufacturing situations.

This book bridges the gap between theory and practice by concentrating on the actual practices used by manufacturing engineers who often have the responsibility of choosing the most economical method of fabrication. Thus, this book gives the engineer the information on adhesives that he needs for comparing adhesive joining with conventional methods of joining; and if the use of adhesives is the answer to the joining problem, the book will help him decide the adhesive and the method of application that will best serve the needs of production. Or, if adhesives are now being used in production, this book will give him the information necessary to determine if the adhesives in use in the plant along with their methods of application are the most economical.

E. J. Bruno

Dearborn, Michigan
February, 1970

ACKNOWLEDGMENTS

In addition to the co-authors listed on pages ii and iii, SME wishes to express its appreciation to E. C. Kron, Chief Metallurgist, Doehler-Jarvis Division, National Lead Company, Toledo, Ohio; Jacob Lichman, Key Polymer Corporation, Lawrence, Massachusetts; and Roy Thom, Vice President, Midland Coating and Chemical Corporation, Troy, Michigan, for their time in planning the book and reviewing the final manuscript.

We also wish to thank the following reviewers: Charles V. Cagle, Project Engineer, The Epoxylite Corporation, South El Monte, California; W. C. Hitt, Manager NDT Systems, Santa Clara, California; J. R. Hubbard, Vice President, Production, Peter Cooper Corporations, Gowanda, New York; and Dr. Charlie Baranauckas, Director of New Product Development, Ray Getty, Department Head, Epoxy Laboratory, and Curt Seyen, Director of Technical Services, H. B. Fuller Company, St. Paul, Minnesota.

TABLE OF CONTENTS

CHAPTER 1 TYPES OF ADHESIVES 1
Chemically Reactive Types 1
Evaporation or Diffusion Types 20
Hot Melt Types (Cooling) 27
Delayed-Tack Adhesives 35
Film Adhesives 36
Pressure Sensitive Types 40
New Technology 41

CHAPTER 2 DESIGN CONSIDERATIONS 45
Advantages and Disadvantages of Adhesive Bonding 46
Materials to be Bonded 47
Area Size to be Bonded 49
Joints 49
Performance Requirements of the Joint 55

CHAPTER 3 SELECTION OF ADHESIVES 59
Physical Requirements 60
Method of Application 63
Selection of Equipment 64
Unit Cost Considerations 66
Plant Facilities 68

CHAPTER 4 SURFACE PREPARATION FOR BONDING 89
Surface Preparation for Metals 89
Surface Preparation for Nonmetallic Materials 90
Abrasive Cleaning 90
Chemical Cleaning 96
Miscellaneous Cleaning Methods 101

CHAPTER 5 METHODS OF BONDING 105
Manual Adhesive Application Methods 105
Machine Application Methods 108

Spray Application Equipment 115
Methods of Completing Assembly 119
Bonding with Hot Melt Adhesives 121

CHAPTER 6 INSPECTION AND QUALITY CONTROL 123
Process Control 123
Bond Inspection and Testing 126

CHAPTER 7 CASE HISTORIES 143
Structural Bonding of Inner to Outer Hood and Trunk Compartment Panels in the Automotive Industry 143
Bonding in the Boat Building Industry 147
Food Packaging: Hot Melt Adhesives 152
Polysulfide Glass Bonding for Automotive Use 155
Aircraft-Bonding of Sandwich Panels 157

GLOSSARY 161

BIBLIOGRAPHY 169

INDEX 177

LIST OF TABLES

Table I-1	Strength Values for a Typical Epoxy Adhesive	**1**
Table I-2	Typical Hardener Systems for Epoxy Adhesives	**5**
Table I-3	Epoxy-Alloy Structural Adhesives	**8**
Table I-4	Unsaturated and Saturated Polyester Adhesives and Use	**9**
Table I-5	Polysulfide-Epoxy Adhesive Properties	**10**
Table I-6	Comparison of a Typical Urethane Adhesive on Aluminum vs Other Adhesives at −400° F	**12**
Table I-7	Role of Silicones in Resin-Glass Laminates	**13**
Table I-8	Comparison between Benzimidazole-to-Metal Shear Strength Bonds at Room and Elevated Temperature	**16**
Table I-9	Properties of a Cured One-Component Epoxy Polymer	**17**
Table I-10	Comparison of Epoxy-Nylon to Other Structural Adhesives	**18**
Table I-11	General Comparison of Hot Melt Adhesives	**29**
Table I-12	Peel Values of Ethylene-Based Adhesives	**30**
Table I-13	Effectiveness of Automotive Safety Glass Adhesives Using PVME Additive to Promote Moisture Resistance	**34**
Table III-1	Examples of Structural Adhesives—Aluminum Bonding	**60**
Table III-2	Adhesive Equipment Application Considerations	**66**
Table IV-1	Maximum Profile Height for Various Abrasives	**93**
Table VI-1	Time Interval Limit for Adhesive Coating of Prepared Surfaces	**125**
Table VI-2	Standard Test Exposures	**140**
Table VII-1	Typical Sealant Requirements on an Average 26 foot Sea Skiff	**151**
Table VII-2	Comparison of Costs Using Temporary vs Permanent Sealant in Sealing a Lapstrake Hull	**151**

ADHESIVES IN MODERN MANUFACTURING

chapter 1

types of adhesives

Currently there are about 25 basic types of synthetic adhesives useful to the design engineer. New types are appearing at an average rate of one per year. Each of these may have from 10 to 20 modifications, and when we consider that frequently two or more individual adhesives are combined to provide special properties, this leads to a vast array of specific products from which to choose. This chapter divides the adhesives into six major types: (1) chemically reactive types, (2) evaporation or diffusion types, (3) hot melt types (cooling), (4) delayed-tack adhesives, (5) film adhesives, and (6) pressure sensitive types. The chapter will give a descriptive evaluation of each type and place it in its proper category so that the reader can narrow his choice of adhesive for a specific application.

A history of the development of our present-day adhesives is outlined in Fig. 1–1.

CHEMICALLY REACTIVE TYPES

Catalytic Plural Components—Chemical Cure

Adhesives of the catalytic plural component class—chemical cure are normally provided in twin tubes or cans and are mixed immediately before use (Fig. 1–2). In large operations, the components may be mixed in bulk to supply an 8-hour shift or mixed directly in the applicator nozzle. Chemically reactive adhesives will harden at room temperature, although the application of heat gives a better quality bond and a faster cure (see Table I–1).

Table I–1. Strength Values for a Typical Epoxy Adhesive.

Property	*Value*
Polymer Tensile Strength	6000 psi
Lap Shear Strength (aluminum)	3200 psi—7 day cure at room temperature 4000 psi—1 hr cure at 176°F(80°C)
Peel Strength, Climbing Drum Test	6 to 12 psi

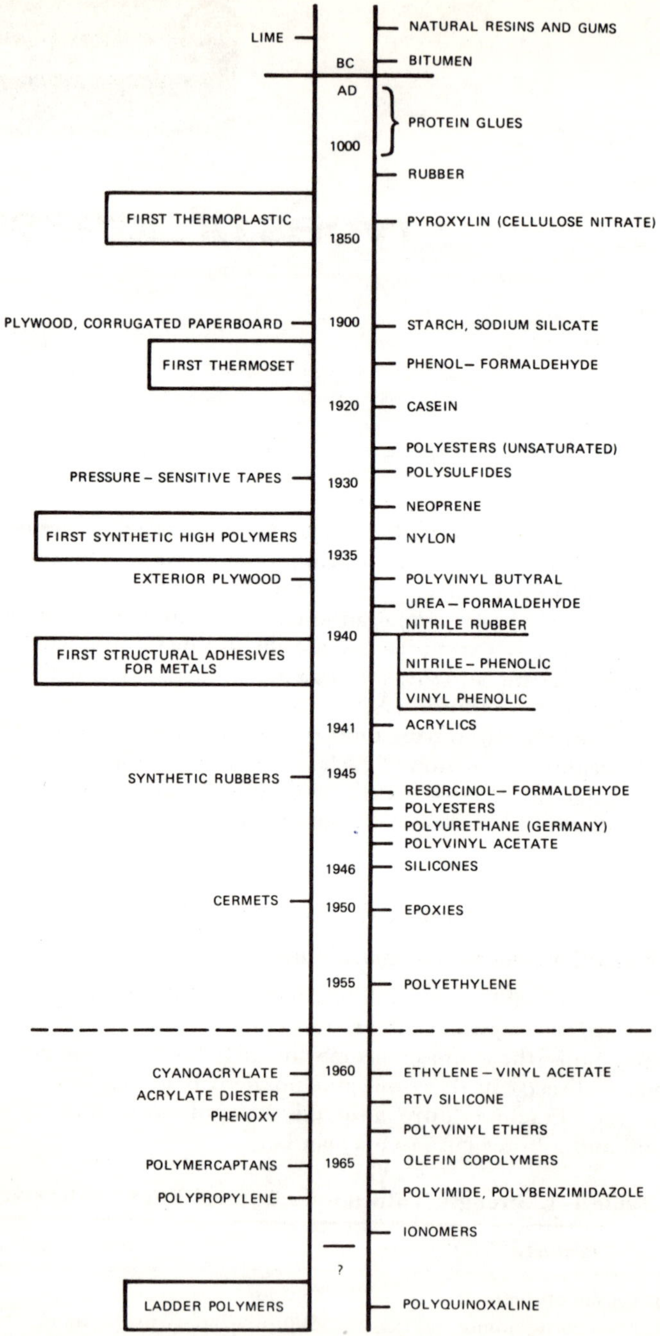

Fig. 1–1. Development of present-day adhesives.

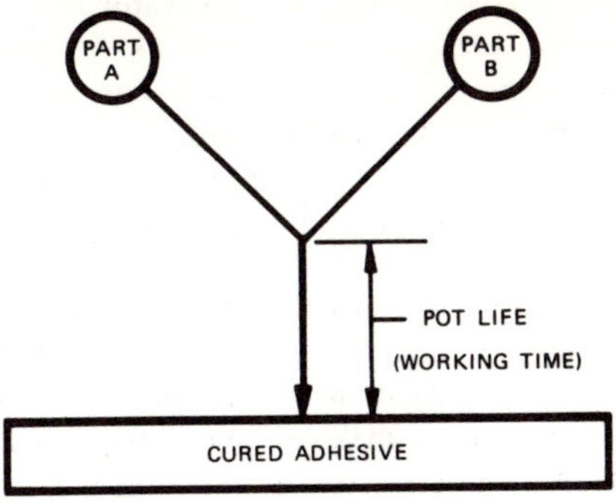

Fig. 1–2. Two-part adhesive.

Epoxy. Epoxy adhesives (general formula below) are familiar to the reader

in the form of twin tube household adhesive kits. They are popular for two reasons: they are relatively simple to use, and a reliable bond is produced between many materials requiring repair or bonding.

Industrial epoxy adhesives may be similar to the household adhesive, or they could be quite different in nature (1). Versatility is an important advantage in the use of epoxies, so the user will find many individual formulations, designed to fit the curing conditions and bond requirements for any of a myriad of specific applications. Epoxies are able to form useful bonds with most materials, and they are particularly useful in joining dissimilar materials. Special attention should be given to certain plastics (polyethylene, silicones, plasticized vinyls, untreated fluorinated polymers) when bonding the epoxies to them. Epoxies are usually chosen for specialized service where high strength is required and a moderately high cost is allowed. Peel strength and flexibility are low for unmodified epoxies.

The most common epoxy resin (general formula below) is represented by the structure:

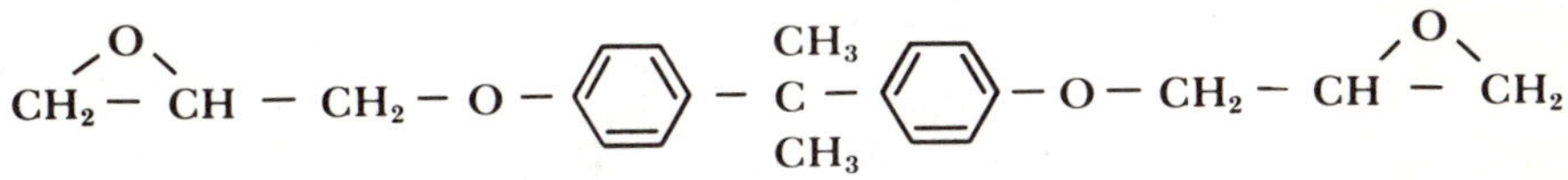

Diglycidyl ether of bisphenol A

Other epoxies include:

1) Higher mol wt homologues of the structure above (more flexible, higher viscosity or solid)
2) Epoxidized polyglycols (flexibilizing additives, adhesion promoters)
3) Epoxidized novolac (high temperature performance)
4) Cycloaliphatic epoxies (little used in adhesives)

5) Epoxidized aliphatics (new, solubility in water may prove useful)

To form a permanent adhesive bond, the above epoxies must be reacted with a hardener (other terminology: catalyst, curing agent, or "Part B"). The hardener chemically converts the liquid epoxy resin into a solid resin. If the epoxy is in a solid form, it must be heated to melt and be able to wet the surfaces being bonded. Generally, only resins that are liquid at room temperatures are used for adhesives.

The epoxy bond will set or cure faster if the bonded surfaces are heated. The most important advantage of the epoxy class of adhesives is that they solidify without evolution of volatile materials and without appreciable shrinkage. Materials with this property may serve as "gap-filling" adhesives.

The major classes of hardeners for epoxy resins are shown in Table I–2. In the catalytic plural component systems, resin and hardener are packaged separately and thus are stable until mixed. The user mixes the components just prior to application. Table I–2 also gives representative curing times and pot life.

Procedures for testing the properties of the cured adhesive bond are discussed in Chapter 6. Some typical test values for a two-part epoxy-polyamide adhesive are given in Table I–1.*

The following list of uses demonstrates the versatility of epoxy adhesives to bond a variety of substrates:

1) Metal—Automobile repair, lamination of (oily) steel sheet and foil, joining of steel pipe (4), underwater protective coatings (5)
2) Concrete—Bonding new concrete to old, patching of cracks (gap-filling)
3) Glass—TV tube safety lamination, stained glass window assembly, and filament wound structures
4) Plastic—Repair of polyester-fiberglass boats, bonding of treated nylon, joining of plastic pipe
5) Wood—Wood to metal bonding (doors, architectural panels, etc.)
6) Ceramic—Binder for ceramic powders (molded, machined, then fired at 1725°C)
7) Composites—Metal-plastic sandwich construction, cement board

Several applications of epoxy adhesives are demonstrated in Fig. 1–3. Epoxy alloys are very important as high strength structural adhesives† for metals. Properties of some epoxy-alloy structural adhesives are shown in Table I–3.

Phenolics. Phenol-formaldehyde resins (general formula below) were the

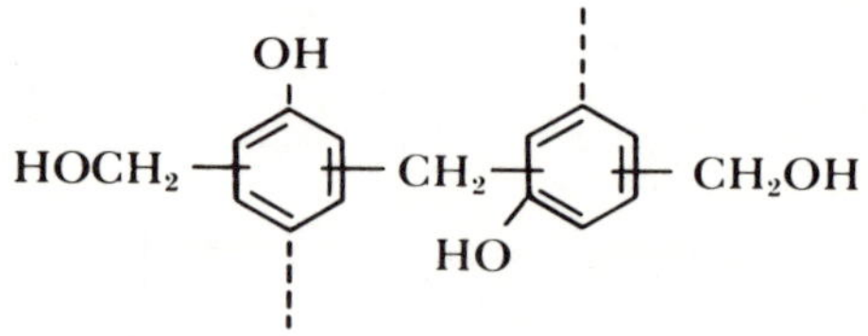

*For additional strength data, see references (2) and (3).

†A *structural adhesive* is usually defined as "an adhesive capable of bonding two materials while transmitting part or all of the stresses between the two members." *Alloying* principles are discussed in this chapter under Chemically Reactive Types and Film Adhesives.

Table I–2. Typical Hardener Systems for Epoxy Adhesives.*

Class	*Chemical Type*	*Cure Time*	*Pot Life (Working Time)*	*Characteristics*
Rapid Setting	Aromatic sulfonic acid	2 to 5 min	1 to 2 min	"Instant-set" adhesive low bond strength
Fast Room Temperature Cure	Primary amine	2 to 4 hrs	30 min	Useful in laminating bonding large surface areas
Slow Room Temperature Cure	Polyamide	8 hrs	4 hrs	Most commonly used epoxy hardener. Need not be proportioned accurately
Fast Elevated Temperature Cure	1) Aromatic amine	2 hrs at 248°F (120°C)	12 hrs	Good bond strengths and improved elevated temperature performance
	2) Tertiary amine	2 hrs at 248°F (120°C)	2 to 24 hrs	Specialized uses – promotes adhesion
	3) BF_3 adducts (blocked lewis acid)	30 min at 248°F (120°C)	<6 months	Semi-latent type hardener
Slow Elevated Temperature Cure	Acid anhydride with accelerator	4 hrs at 302°F (150°C)	~1 week	Good elevated temperature properties

*Check manufacturer's literature for correct proportions of resin and hardener.

Bonding old to new concrete

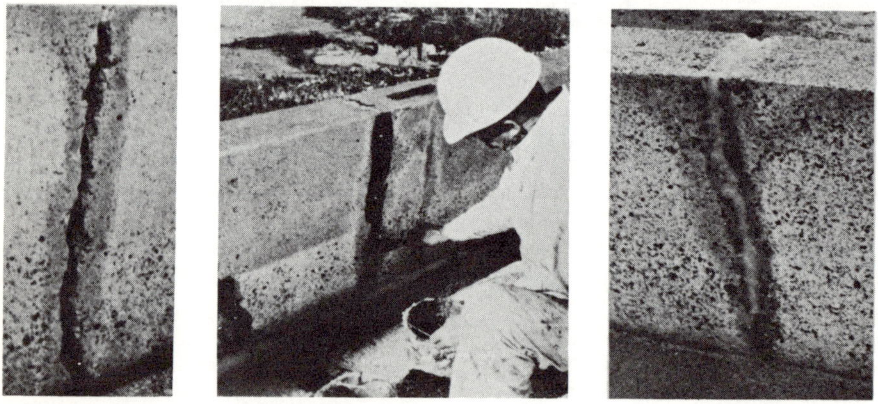

Crack-filling in old concrete

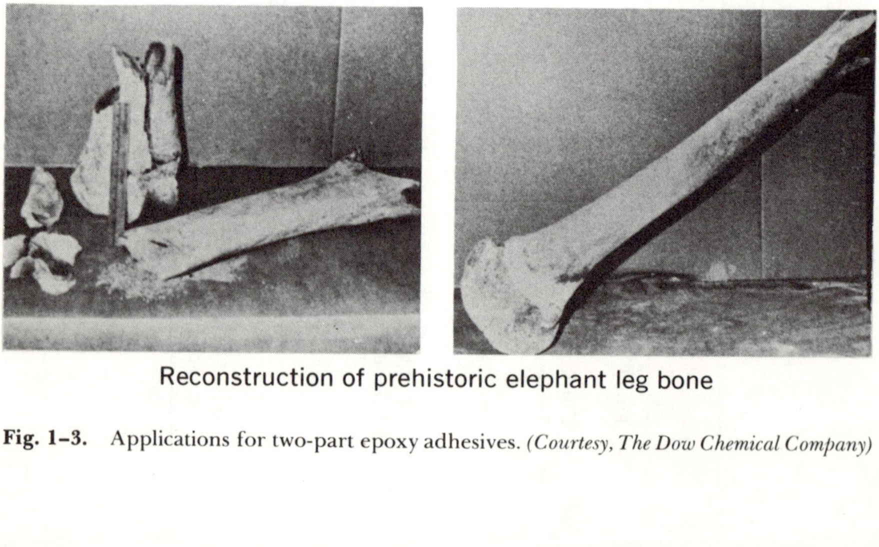

Reconstruction of prehistoric elephant leg bone

Fig. 1–3. Applications for two-part epoxy adhesives. *(Courtesy, The Dow Chemical Company)* earliest synthetic plastics to attain commercial prominence (6). Although they are mainly used primarily to mold plastic products, they also dominate the wood adhesives (plywood) field. For this reason, phenolics represent the largest volume of any synthetic adhesive, at a current usage of 500 million lbs/yr. They are also among the lowest cost adhesives.

Phenolics may be formulated as water dispersions which have advantages in bonding wood where penetration into the cell structure is vital for the formation of permanent bonds. Since phenolics have a high temperature stability, they are used in blends with other resins ("alloys") for bonding to metals. These bonds are exceptionally stable to their environment (moisture, temperature, microorganisms, etc.). Limited shelf stability before usage and the release of volatiles during cure are potential disadvantages.

Beyond the wood and wood products area, the uses for unmodified phenolic adhesives are mainly: (1) *primers*, to prepare metal surfaces for bonding, and (2) *binders*, for such varied products as glass wool insulation mats, foundry sand, abrasive wheels, and brake lining composites.

Phenolics are provided either as a one-component heat-curable liquid solution or powder, or as a liquid solution to which a catalyst must be added. The curing mechanism differs for these two types, heat curing and catalyst curing, as shown in the general formula below:

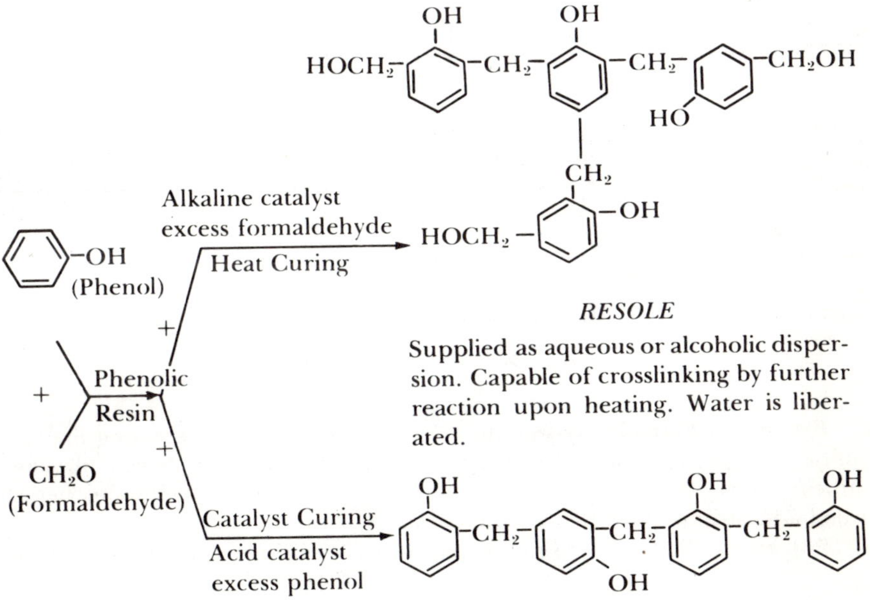

RESOLE

Supplied as aqueous or alcoholic dispersion. Capable of crosslinking by further reaction upon heating. Water is liberated.

NOVOLAC

Requires addition of a hardener (hexamethylene tetramine) to produce a thermoset resin.

Table I–3. Epoxy-Alloy Structural Adhesives.

Adhesive System	*Typical Tensile Shear Strength at Room Temperature ASTM D-1002*	*Typical Peel Strength at Room Temperature ASTM D-1781*
	(psi)	*(lb/in)*
Epoxy-Nylon	6500	90
Epoxy-Phenolic	3200	4
Epoxy-Polyamide	4000	12
Epoxy-Polysulfide	3000	20
Epoxy-Urethane	5200	90
MIL-A-005090 Metal-to-Metal, Class 1 Specifications (superseded by MMM-A-132)	4500	50

Phenolic Alloys. To improve the usefulness of phenolics as structural adhesives, they may be alloyed with a second component such as a thermoplastic or an elastomer (Fig. 1–4). The blend or alloy utilizes the structural advantages of each component to produce a stronger or tougher material (7).

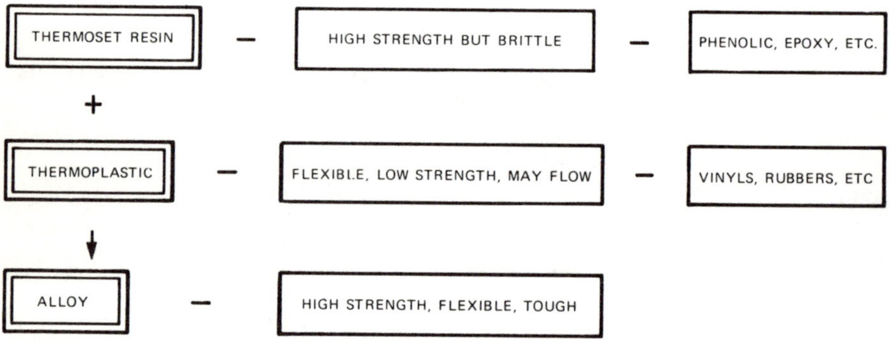

Fig. 1–4. Alloying parameters.

Resorcinol-Formaldehyde. This related resin is used in tiecoating tire cord fabric to rubber (8), and is also noted as an extremely durable "marine plywood" adhesive. It is curable at room temperature in 8 to 12 hours whereas the phenolic wood adhesives require elevated temperature cure.

Polyesters. Polyester resins (general formula below) may be divided into

$$\sim R - CH_2 - O - \overset{\overset{\Large O}{\|}}{C} - R' \sim$$

two distinct groups (see Table 1–4) (9): saturated and unsaturated.

The saturated polyesters are reaction products of difunctional acids plus difunctional alcohols or glycols. The adhesive uses for the saturated polyesters are minor and will not be discussed.

Table I–4. Unsaturated and Saturated Polyester Adhesives and Use.

Type of Adhesive Unsaturated	*Use*	*Type of Adhesive Saturated*	*Use*
Polyesters	Glass laminates	Polyethylene-terephthalate	Synthetic fibers Films Molding plastics
Alkyds	Alkyd coatings	Polycarbonate	Hot melts Solution adhesives
Vinyl Esters	Minor adhesive uses		

The unsaturated polyesters which require a catalytic cure have a few uses as adhesives. These usually involve the bonding of polyester substrates. Uses do include polyester patching kits for repair of fiberglass boats, automobile bodies, and concrete flooring. Other minor uses are in bonding polyester laminates to polyester or to metal, and as an adhesive for optical equipment. For the majority of other applications, where the properties of adhesion, low shrinkage, and solvent resistance are required (10), a different type of adhesive, such as epoxy, will be chosen.

An example of the chemistry is as follows:

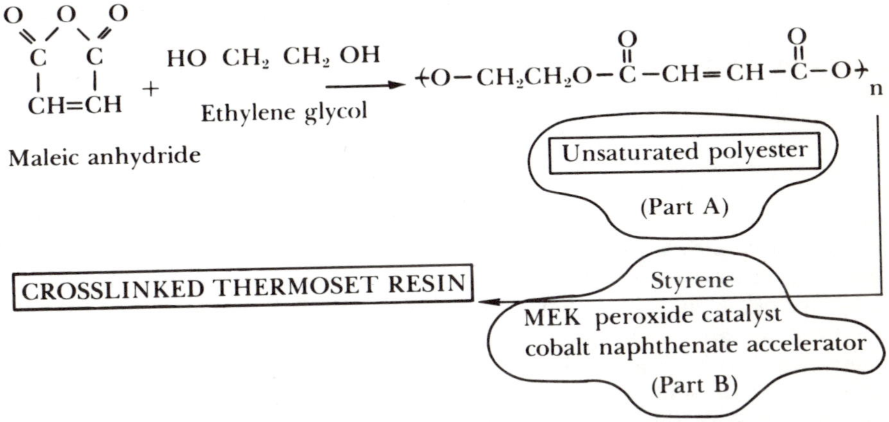

The shelf life of the inhibited intermediate is over one year. With cobalt accelerator, the adhesive mix will set in 4 to 8 hrs at room temperature. The purpose of the styrene is two-fold; it is both a solvent for and a coreactant with the polyester.

Polysulfides. Polysulfides (general formula below) are flexible materials

$$HS \sim\sim S - S \sim\sim SH$$

that belong to the synthetic rubber family. Although they are principally used as sealants for automotive, construction, and marine fabrications (11), they have some utility as flexibilizing hardeners for epoxy adhesives. The sulfur linkages combine good strength with the ability to rotate freely, thus giving a strong, flexible polymer. Polysulfides and epoxies are mutually soluble in all proportions. As the percentage of polysulfide increases, the cured product changes in properties from an adhesive to a sealant. They are also often used to bond new to old concrete. In addition, steel-glass bonds using this combination have good properties, but the cost is relatively high (12). For adhesive sealants, curing agents such as lead dioxide* or an isocyanate* are used. Among other techniques, a three-component system is available that is mixed directly in the applicator head for metal to glass bonding.

The new polymercaptans and mercaptan terminated nitrile polymers give properties similar to the well established polysulfides but with differences in reactivity and chemical resistance. Any of these may be used as one part of a two-component adhesive with a filled epoxy resin as the other. The adhesive properties for a 1:1 ratio of the two polymers are shown in Table I–5.

Table I–5. Polysulfide-Epoxy Adhesive Properties.

Property	*Value*
Pot Life	30 min
Cure Time	24 hrs (with tertiary amine accelerator)
Lap Shear Strength	600 psi, steel-to-wood or steel-to-glass
Peel Strength	12 to 20 lb/in width. steel-to-steel

Polyurethanes. Urethane polymers (general formula below), developed in

$$\sim\!\sim R - \overset{\displaystyle H}{N} - \overset{\displaystyle \overset{O}{\|}}{C} - O - R'\sim\!\sim$$

Germany during World War II, have found a place in flexible foams, cryogenic sealants, and abrasion-resistant coatings (13). As adhesives, the principal use is bonding "difficult-to-bond" plastics, usually to a dissimilar material or to metals. The use of polyurethanes for adhesives is expanding but is still quite small.

Urethane adhesives bear a resemblance to epoxies in three respects: cost, method of handling, and ability to bond to most surfaces. The difference from epoxy may best be illustrated by considering the cured urethane as a very lightly crosslinked thermoset resin—almost a thermoplastic. This gives a flexible, rubbery characteristic in contrast to the hard and rigid epoxies. This flexibility combined with good adhesion indicates good bonding to flexible plastics, where peel strength is important (14). The best feature is strength at cryogenic temperature which is better than for any other adhesive. Table I–6 compares strength at −400°F.

A potential disadvantage is the reactivity of the uncured adhesive with moisture, which may impose some handling precautions. This reactivity, however, introduces the concept of a one-component moisture curing adhesive (15 and 16). Poor lap shear strength at room or elevated temperature and sensitivity of the cured glue lines to moisture are other limitations. The lap shear strength

*Both are toxic chemicals. Safety considerations should be carefully studied. Check safety instructions from the manufacturer. See also Chapter 3.

for a typical urethane adhesive with aluminum metal is only 1800 psi at room temperature, but this increases to 5000 psi at −100°F and below.

The chemistry is complex due to the variety of principal and side reactions that may occur. Examples of the desired reactions are:

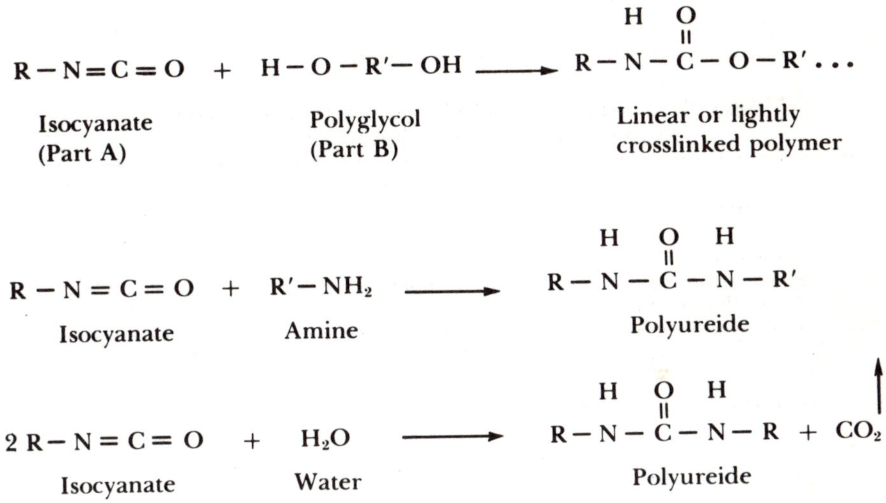

Both components must be di- or polyfunctional to build up the molecular weight of the polymer and provide crosslinking.

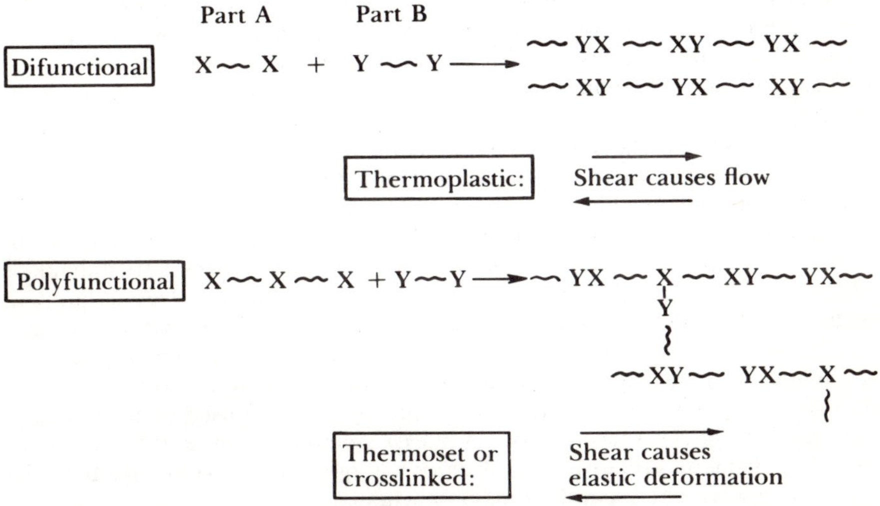

Specific uses for urethane adhesives include the following:

1) Rubber to metal—Lap shear strengths of 1000 psi are reported
2) Packaging film lamination—Polyester-to-polyethylene laminates use a two-part urethane adhesive. This will be described under solution adhesives.

3) Flocking adhesives—The bonding of short fibers (synthetic) to fabric by an electrostatic technique. This is another example of the use of a flexible adhesive to bond dissimilar plastics.
4) Laminating of flexible vinyls to nylon or other plastics—This is a different problem for which urethanes work well (17).
5) Tire cord bonding—Urethanes may be used to promote adhesion of both nylon and polyester fiber to rubber.

Table I–6. Comparison of a Typical Urethane Adhesive on Aluminum vs Other Adhesives at −400°F.*

Adhesive	*Lap Shear Strength (psi)*	*Peel Strength (lb/in)*
Urethane	8000	26
Epoxy-Nylon	4600	(brittle)
Epoxy-Polyamide	1600	(brittle)

*See Table I–10 for a comparison over a broad temperature range.

Silicones. Silicone polymers (general formula below) have unusual proper-

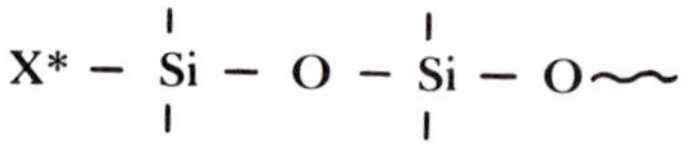

ties and are used both to promote adhesion and to prevent adhesion. Commercial products such as silicone greases, rubbers, and fluids take advantage of their good heat stability and chemical inertness. Their surface active properties make them useful as defoamers, water repellants, and mold release agents. Adhesive uses fall into four groups: (1) *primers* or coupling agents, used to solve some difficult industrial bonding problems, (2) *adhesives* and *sealants*, (3) *pressure-sensitive* adhesives, and (4) *heat-cured* adhesives (18).

Usage of silicones as adhesives is limited mainly by the high cost. Applications are numerous and varied. In general, silicones are used where organic materials cannot withstand harsh environmental conditions, where superior reliability is required, or where the durability of silicones gives them economic advantages. As coupling agents, silicones are widely used for surface treatments of fiberglass fabric for glass-reinforced laminates. Epoxy or polyester bonding to glass cloth treated with a silicone "finishing agent" is improved both in strength and in moisture resistance of the cured bond (Table I–7) (19). The retention of flexibility and some degree of strength over a temperature range from cryogenic to over 500°F is unusual in the field of polymers, and is used to advantage in sealants for the aerospace industry (20). In general, however, the room temperature strength properties of silicones are quite low in comparison to the typical polymer. The formulas for the reactions follow:

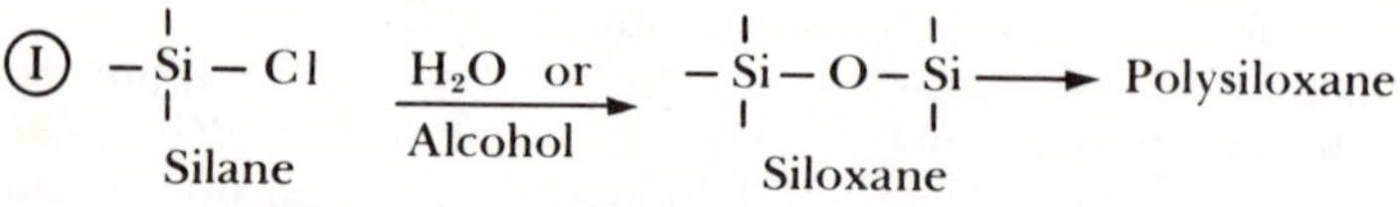

*X represents OH, CH_3, or some organic functional group.

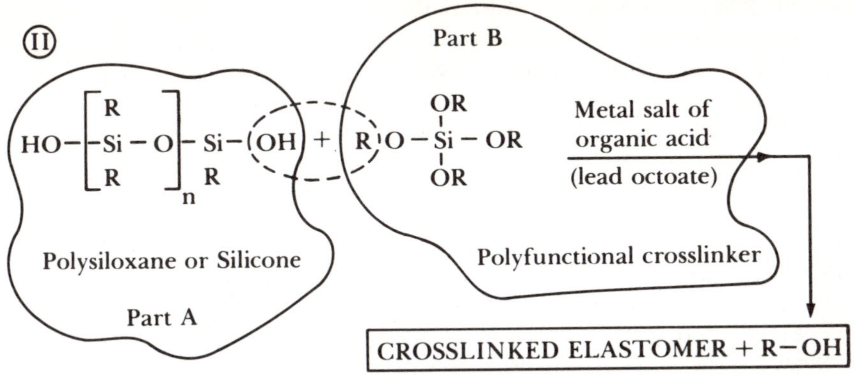

Joints bonded with silicones are designed to utilize the good peel strength of silicones rather than the tensile or lap shear properties. This is discussed in Chapter 2. For example: the peel strength in bonding rubber to aluminum is from 17 to 20 lb/in; the peel strength for urethane sealant to aluminum is: (1) 3.5 lb/in without primer and (2) 14 lb/in with a silicone coupling agent. The lap shear strength, metal-to-metal, is 250 to 500 psi.

Specific adhesive application of silicones include:

1) Two-part adhesive for bonding insulating tape to magnet wire (Class H performance)
2) One or two-part adhesive to prepare pressure-sensitive tapes useful from −80°F to +500°F. These are used for adhering certain components in electronics or in aerospace assemblies.

The use of silicone as a primer includes:

1) Bond promoters with phenolic binders for foundry sand or abrasive wheels
2) Filler treatment in filled polyester or epoxy castings (epoxy concrete patching formulations)
3) Improved bonding of polysulfide or urethane sealants to metal substrates or to glass.

In some cases the silicone is equally effective when blended into an adhesive formulation as when it is applied separately as a primer (Table I–7).

Table I–7. Role of Silicones in Resin-Glass Laminates (19).*

Role	Silicone Percent	Flexural Strength Dry (psi)	Flexural Strength 2 hr Boil (psi)	Compressive Strength Dry (psi)	Compressive Strength 2 hr Boil (psi)
	0	62,600	36,300	30,300	7,100
As "Finish" on Glass Fabric	.16	77,000	70,000	40,600	33,800
As Additive to the Resin	.2	68,500	62,700	36,800	21,100

*The data are for a polyester resin.

Catalytic Plural Components—Moisture Cure

For this curing mechanism, the adhesive is supplied as a one-component formulation, usually in a collapsible tube to insure good protection from atmospheric moisture before use. The second component, moisture, is always available, either from the surrounding air or as adsorbed moisture on the surface of the substrates to be bonded. The only situation where insufficient moisture may be present is in overlap bonding of non-porous substrates such as metals.

Silicones. Organic groups such as acetyl, attached to a silicon atom, are readily hydrolyzed. "Room temperature vulcanizing" (RTV) sealants use the principle in which the adhesive reacts with atmospheric moisture and releases acetic acid vapor, causing a crosslinking of the viscous paste into an elastomer (21), exemplified by the silicone rubber adhesives.

For silicone coupling agents, moisture adsorbed on the substrate plays an important role in attaching the silicone molecule through hydrolysis. The opposite end of the molecule contains a chemical group such as a vinyl or amine which is reactive with epoxy, polyester, or other resin that is to be adhered to the substrate. In this way, a single layer of silicone molecules couples the resin to the substrate. Along with bond strength improvement, improved moisture resistance usually results.

Urethanes. The free isocyanate groups in a urethane adhesive will react with water. This serves as a means of converting the liquid into a cross-linked solid. This technique is used in coatings and sealants. Adhesive use is minor because of slow curing, foaming during cure, and variable shelf life.

Other Adhesives. Polysulfides utilize atmospheric moisture to accelerate cures. However, a two-component system is usually used, consisting of formulated polysulfide and formulated lead dioxide. Moisture converts a portion of the latter to a faster reacting form.

Cyanoacrylate adhesives cure very rapidly through surface moisture and may be used as quick-set cements to bond metal to non-metal. Lap shear strength of 2000 psi has been reported; however, its resistance to moisture is still somewhat low.

A one-container epoxy has been developed which utilizes atmospheric moisture to effect the cure by an unusual mechanism. The amine hardener for the epoxy is first adsorbed by molecular sieves. The sieves, in a bead form, are then blended into the epoxy formulation. The sieve structure is a porous inorganic matrix that prevents contact of the trapped amine with the surrounding epoxy. Water molecules are preferentially adsorbed by the molecular sieves and will displace the amine from the pores. Thus upon usage, exposure of the adhesive to atmospheric moisture causes the amine hardener to be released into the epoxy and the system will cure. Alternately, heating the glue line will also release the epoxy hardener, and thus this system may be used as a one-component heat-activated epoxy adhesive. One-component urethanes may be formulated using this same principle. A tertiary amine is the hardener.

Heat Activated (One-Part System)

One-part adhesives are attractive because they eliminate the need for careful proportioning and mixing of components. They may be supplied as solid films (see Film Adhesives, below) for convenient assembly of the joint and a uniform bond line thickness. These systems do require elevated temperature cure and may have a limited shelf life.

Benzimidazoles. Benzimidazole polymers (general formula below) were

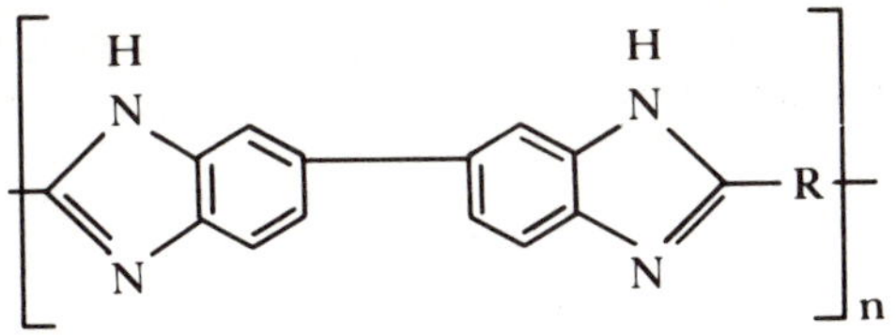

R is preferably aromatic

developed as a result of research specifically aimed at producing high temperature organic polymers. Polyimides are analogous polymers competing for the same applications (22). Benzimidazole polymers are thermoplastics, although the thermoplastic nature is not evident below 700°F. They are relatively stable in air to 550°F (300°C) for short-term applications (see Fig. 1–5) (23). Polyimides may be favored for long-term strength retention.

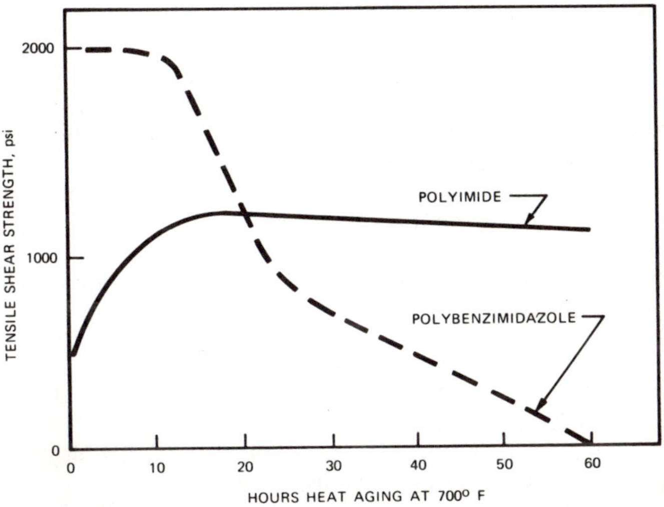

Fig. 1–5. High temperature adhesive performance (23).

These materials are just emerging from the developmental stage. Improvements in properties and in workability may be expected. The formulations usually contain oxidation inhibitors and a filler (aluminum powder). The adhesive prepolymer cures by a high-temperature condensation reaction involving the loss of water or other volatiles.

Products cost very high and are limited to the bonding of high-temperature metals (stainless, titanium) that must perform at 500°F or higher (24 and 25). Benzimidazoles are of most interest to the aerospace engineer for use in the adhesive assembly of lightweight honeycomb structures for supersonic aircraft, missiles, and other space systems (26). Some moisture sensitivity at room temperature is indicated in the data in Table I–8. These properties drop gradually upon heating to 600°F, then more rapidly above this temperature.

Table I–8. Comparison Between Benzimidazole-to-Metal Shear Strength Bonds at Room and Elevated Temperature.

	Lap Shear Strength	
Temperature	*Titanium (psi)*	*Stainless (psi)*
Ambient	3500	3700 1900 (after 1 week in water at 160°F)
600°F	2750	2650
1000°F	1400	1400

Peel strength: 49 lb/in at room temperature (climbing drum test)

Epoxies. Hardener systems for epoxies are listed in Table I–2. If the hardener is a meltable solid rather than a liquid, it may, in some cases, be included in a formulation supplied to the user as a latent "one-component" adhesive. Dicyandiamide (DICY) is the standard hardener used in this way, and requires a cure temperature of 350°F or higher. Recently, accelerators for DICY have been found that reduce the cure schedule to one hour at 250°F (27). This is more practical for the bonding of large surface areas where the supply of heat is limited. The boron trifluoride-amine complexes are second to DICY as hardeners in one-component epoxies. They are neither as latent nor as effective in curing as DICY—final bond strengths are lower—but may fit a given cure schedule to better advantage.

A typical one-component paste-form epoxy adhesive supplied to the user will include the following:

1) Principal epoxy resin
2) Hardener—accelerator combination
3) Thixotroping agent. This prevents excessive flow of the resin when heated to the cure temperature
4) Filler. Aluminum powder of 150 mesh is customary in metal-to-metal adhesives, but other fillers can be also used
5) Additional flow-control agents
6) Special epoxy resin additives. These promote specific properties such as reactivity, high temperature strength, adhesion, etc.
7) Modifier, usually a thermoplastic. This eliminates brittleness in the cured polymer (see Table I–3).

Physical properties of a cured epoxy polymer of this type are given in Table I–9 (28). Fig. 1–6 is a cross sectional view of this product at 30,000 magnification.

Major uses for the heat-activated epoxies include:

1) Replacement of lead solder or rivets in metal joints (29)
2) Sandwich panel adhesive—bonding of skins to core in heavy-duty, lightweight panels

3) Lightweight honeycomb structure assembly for aircraft
4) Replacement for welding in pressure-tight assemblies.

In all of these, the filleting action of the adhesive formulation is important in producing a strong, thoroughly bonded joint. The situation is analogous to

Table I–9. Properties of a Cured One-Component Epoxy Polymer (28).

Property	*Value*
Tensile Strength	11,000 psi
Shear Strength	7,000 psi
Flexural Strength	13,000 psi
Modulus of Elasticity	2.5×10^5 psi
Coeff of Thermal Expansion	60×10^{-6} in/in/ °C
Heat Deflection Temperature	240°F

Legend: Resin = Diglycidyl ether of Bisphenol A
Hardener = DICY
Filler = Aluminum powder
Modifier = Nitrile rubber
Cure = 30 min at 360°F

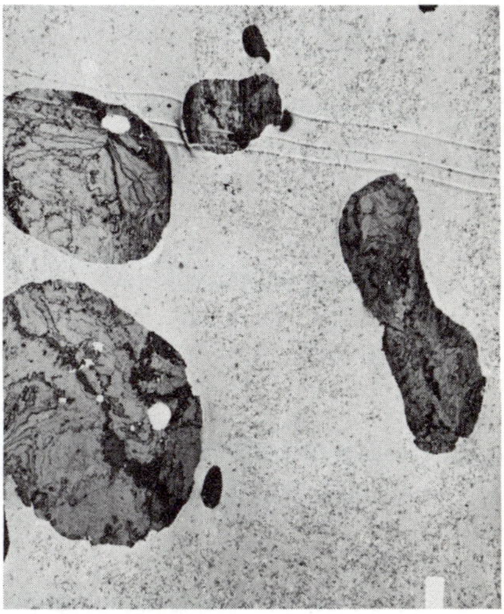

Fig. 1–6. Cross sectional view of a cured one-component epoxy adhesive at 30,000× magnification. Aluminum filler and finely dispersed collodial silica gellant are prominent. This view shows the typical outline and separation between aluminum particles (the large dark shapes) and also the "salt and pepper" appearance of the properly distributed gellant. The epoxy itself is transparent. Notch width = 1 μ. *(Courtesy, The Dow Chemical Company)*

the buildup in welding or soldering. The ability of the supplier to formulate and mix properly is also called upon here.

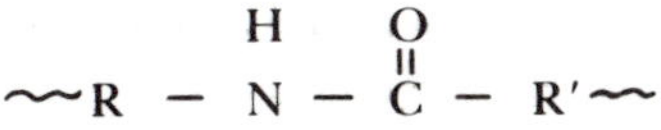

Nylon. Nylons form the high molecular weight members of the family of polyamide resins (30). Although these polymers have excellent physical properties approaching 12,000 psi tensile strength combined with 300 percent maximum elongation, the high softening points and high melt viscosities limit their use as hot melt adhesives. However, certain specialty nylon resins having lower melting temperatures are used quite successfully by extrusion techniques. Both nylon 11 and high molecular weight (mol wt) polyamides resins chemically related to dimer acid-based polyamides are used as high strength metal-to-metal adhesives and are applied by extrusion.

Terpolymers and methyl methoxy modified nylons are soluble in certain organic solvents and can be combined with epoxy resins to form exceptional metal-to-metal adhesives.

Table I–10 compares lap shear strength to other adhesives used for the same purposes (31). Peel strength of these nylon-epoxy combinations also is exceptional; values over 100 lb/in by the climbing drum peel test are possible (32).

Table I–10. Comparison of Epoxy-Nylon to Other Structural Adhesives.

	Test Temperature					
Adhesive	*−423**	*−320**	*−100**	*+77*	*+180*	*+250*
Epoxy-Nylon	4580†	5050	5500	6000	1750	700
Nitrile-Phenolic	1710	4400	5020	2960	2500	2000
Epoxy-Polyamide	1640	1760	1850	2180	1600	(Low)
Polyurethane	8000	4900	4500	1800	1100	700

*(31)
†Values are lap shear strength in psi on aluminum.

A latent hardener for the epoxy (dicyandiamide) is included in the epoxy-nylon blend, although reaction of the epoxy with the nylon is also indicated at the cure temperatures used (33). Epoxies and nylons are not truly compatible at room temperature and must be "forced together" in formulating the adhesive. For this reason the products are supplied primarily as flexible films.

As an alternate to epoxy-nylon blends, a nylon may be used as a powdered filler in an adhesive. As the glue line is cured at high temperature, the filler fuses and contributes its special strength properties to the finished bond. Peel strength in an epoxy/polyamide adhesive may be increased to 40 lb/in by adding 20 parts of nylon powder per hundred parts of resin. See also Table I–3 for comparison with original peel strength values.

Phenolics. Phenolic adhesives have been described in Catalytic Plural Components—Chemical Cure above. Phenolics available as one-component heat activated systems include:

1) Meltable powders of the resole type, used as binders for particle board, etc.
2) Alloys, such as epoxy-phenolic or nitrile-phenolic, used in structural bonding of metals.

Polyvinyl Acetals. These resins (general formula below) are derived from

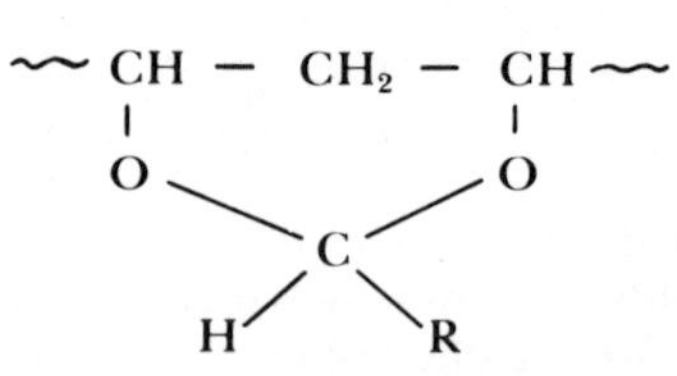

polyvinyl acetate. The acetate is hydrolyzed to polyvinyl alcohol (which is also an adhesive) and then reacted with an aldehyde to form the acetal:

Formaldehyde ⟶	*polyvinyl formal* }	Both blend with phenolics
Acetaldehyde ⟶	*polyvinyl acetal* }	producing structural adhesives.
Butyraldehyde ⟶	*polyvinyl butyral*	Safety glass laminates

Conversion is only about 80 percent. The resulting mixture of reactive groups in the polymer will crosslink upon heating.

Excellent adhesion is the principal advantage (34). This is offset somewhat by poor moisture resistance. Clarity and adhesion to glass has allowed polyvinyl butyral to gain an established position in the manufacture of safety glass. Polyvinyl acetals are useful in flexibilizing thermoset resins to obtain structural adhesives for metals. The vinyl-phenolics and vinyl epoxy alloys are examples. The "Redux" system, developed in Great Britain during World War II, combines granular polyvinyl formal with a liquid phenolic resin. This cures at about 300°F and is useful in aircraft assembly and sandwich construction as a low-cost structural adhesive. Lap shear strength is 5000 psi at room temperature and 1500 psi tested at 250°F. Peel strength is 15 lb/in, which may be increased at the expense of lap shear strength by a higher acetal/phenolic ratio. Polyvinyl acetal serves as a flexible cement for belting, and to some extent as a china cement, wood glue, and in laminated products. These polyvinyl acetals should not be confused with the recent (1960) acetal resins. The latter include poly-formaldehyde and homologues, which are coming into use as molding plastics, but have no adhesive applications.

Rubber Compounds. Soft rubber linings (general formula below) on process

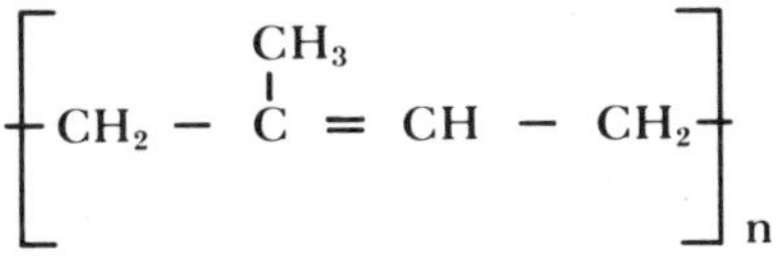

equipment must be bonded to the surfaces to be protected. In some cases the rubber is supplied in rolls as a self-adhering heat-curing prepolymer (35). Therefore it may be considered as a special type of adhesive. Refrigerated storage extends the shelf life. As the product is applied, a solvent may be wiped onto the surface to provide tack. Rubber adheres well to metals and to all other materials except a few plastics, but good adhesion even to these plastics may be accomplished by the use of primers. A varnish of carbon disulfide* is often applied to the exposed surface as a cure accelerator. The system will cure in 3 hrs at 150°F to a soft rubber (55 Durometer A) or in 16 hrs at 190°F to a hard rubber (90 Durometer).

Urethanes. One-component heat curable urethane formulations are used chiefly in sealants. Here the "blocked isocyanate" principle is utilized. The free isocyanate groups are all reacted with a temporary blocking agent such as phenol. This allows the isocyanate prepolymer to be blended with the polyglycol curing agent and packaged as a mixture that is stable until heated to 280 to 300°F. In this reaction, phenol is released, and the isocyanate groups are regenerated. At this temperature the reaction with polyglycol is very rapid to form the cured urethane polymer.

* A toxic material. Safety considerations should be carefully studied. Check safety instructions from the manufacturer. See also Chapter 3.

EVAPORATION OR DIFFUSION TYPES

Solvent-Based Systems

Solvent-based adhesives are supplied by coating the solution on a substrate or substrates followed by some combination of solvent evaporation and contact with the second adherend which may or may not be coated. With porous materials such as wood and paper, the contact usually precedes the final drying step, and the last traces of solvent are absorbed by the substrate. On the other hand, metals and non-porous plastics require essentially complete removal of solvent, so that good contact will be maintained during the bonding process. In the latter case, heat and pressure are used to activate the adhesive coating during bonding. The advantages of solvent systems are good water resistance, high wet strength, and ease of wetting some substrates. The disadvantages of solvents are the fire hazard they present and their toxicity.

Natural Rubber. Natural rubber is one of the earliest raw materials used in the preparation of adhesives. Isolated from a tropical tree (*Hevea brasiliensis*) as a white, watery, fluid latex, the solid rubber product is obtained by coagulation and washing or stripping. The most important property of the crude rubber from an adhesive viewpoint is tack or stickiness. Natural rubber is also soft and flexible with relatively low strength values unless vulcanized (chemically cross-linked). These properties make natural rubber suitable for use in pressure sensitive applications and in bonding nonmetallics such as leather, textiles, and paper.

In 1960, the consumption of natural and synthetic rubber for adhesive applications was 210 million pounds and about half of that amount was natural rubber (36).

Solutions of natural crepe rubber in solvents comprise the standard household "rubber cement." Again, advantage is taken of the initial tack of the rubber. However, natural rubber loses most of this tack on aging. Natural rubber has good water resistance but is susceptible to attack by solvents, oils, and chemicals. Ozone and other oxidizers will gradually reduce flexibility and adhesion. To combat this, antioxidants are generally added to adhesive formulations. A variety of natural and synthetic resins can be blended with natural rubber in solvent-based adhesives. Variations in tack, adhesive strength and other properties can be achieved by formulation.

The chemical structure of natural rubber is best described as a *cis*-poly-

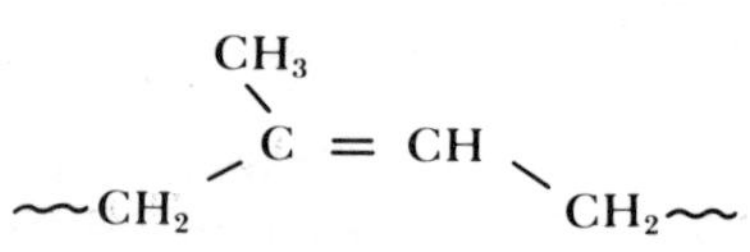

isoprene. The repeat unit as shown here contains a double bond and thus reactive sites are available for reaction with vulcanizing agents such as sulfur.

Important uses of natural rubber solvent-based adhesives are in the building, automotive, and shoe industries. Rubber cements are used for adhering nonmetallic trim to metal automobile parts. An example of a major use in construction is in the preparation of mastics for floor tile. In shoes, natural rubber cements provide water-resistant, flexible bonds to leather, rubber, and canvas.

Reclaimed Rubber. Solvent adhesives based on reclaimed rubber have a number of applications which are similar to the natural rubber systems. Reclaimed rubber is usually obtained from used tires and thus contains small amounts of fibers, sulfur, and other additives. The reclaiming process must, however, meet certain standards of the rubber industry.

Mixtures of natural or synthetic rubber with reclaimed rubber are used in adhesive formulations when a more economical system is required. Although the reclaimed material has relatively low elongation and tensile strength, the economic advantage will sometimes dominate the choice of an adhesive. Modified with resins, tackifiers, and fillers in solvent-based systems, reclaimed rubber will provide a good compromise in physical properties. Mixtures with asphalt are used extensively in the automotive industry for body sealants and numerous adhesive applications. Solvent-based mastics are frequently made from reclaimed stock for adhering rubber or plastic tile to wood subfloors.

Synthetic Rubbers. A number of synthetic, rubber-like materials have become available during the last several decades with properties that are useful in adhesive applications. Most important are nitrile rubber, neoprene, butyl rubber, and styrene-butadiene rubber.

Nitrile rubber is a copolymer of butadiene and acrylonitrile having the following repeating unit:

$$\sim CH_2 - CH = CH - CH_2 - \underset{\underset{CN}{|}}{CH} - CH_2 \sim$$

Usually the copolymer contains enough acrylonitrile (over 25 percent) so that good resistance to oil and grease is obtained. The adhesive properties also increase with increasing nitrile content. Nitrile rubbers are used to bond vinyls, other elastomers, and fabrics where good wear, oil, and water resistance are important. Compatibility with additives, fillers, and other resins is another advantage of nitrile rubber adhesives.

Phenolic resin blends with nitrile rubber have many important applications as structural adhesives. In the aircraft industry, aluminum honeycomb construction is fabricated with nitrile-phenolic adhesives. Other uses include bonding of abrasives to metal, brake linings to brake shoes, and lamination of aluminum and stainless steel. These adhesives are used in the form of supported or unsupported films and then heat-cured in the bonding operation to give optimum bond strength to the substrate. In bonding metals, tensile shear strengths up to 4000 psi at 70°F and 1000 psi at 320°F are achieved (37).

Neoprene or polychloroprene (general formula below) is used extensively

$$\sim CH_2 - CH = \underset{\underset{Cl}{|}}{C} - CH_2 \sim$$

to bond aluminum and is usually compounded and dissolved in organic solvents for convenient application. Although neoprene is similar in properties to natural rubber, it generally provides higher strength, better aging, and higher temperature resistance. Neoprene solvent-based cements are also used extensively as shoe adhesives. For structural applications, neoprene is usually combined with a phenolic resin plus a number of other additives for curing and stabilizing the mixture (see Fig. 1–7). Both cold-setting and heat curing formulations can be prepared.

Butyl rubber is a copolymer of isobutylene with small amounts of isoprene.

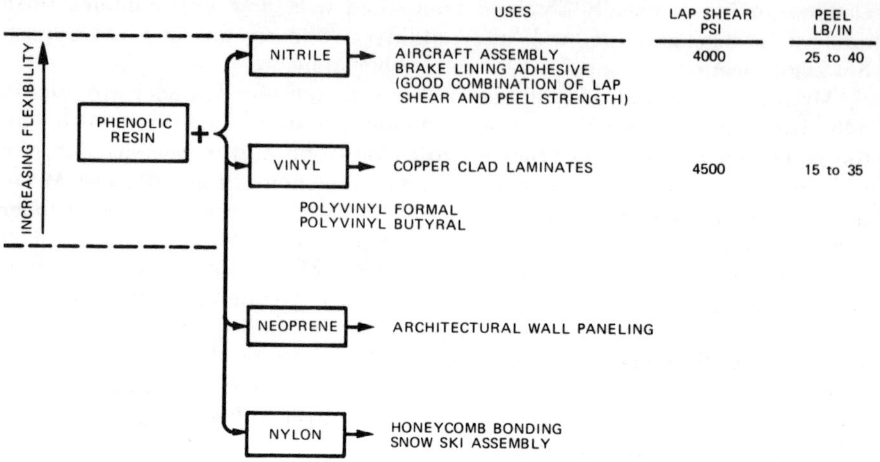

Fig. 1–7. Phenolic alloys.

Characterized by relatively low strength levels and a tendency to creep under load, butyl rubber cements are useful in packaging applications where advantage can be taken of the water resistance and low gas permeability of the rubber. Butyl rubber is also used as an adhesive sealant.

Styrene-butadiene rubber when properly formulated with reinforcing agents and resins, can provide comparable bond strengths to natural rubber. Commonly available as a copolymer composition of 75 percent butadiene and 25 percent styrene, styrene-butadiene rubber (SBR or GRS) was the first synthetic rubber produced on a large scale, and it is the largest volume synthetic rubber manufactured in the U. S. today. Most of the production is used for products other than adhesives, but when a low cost, synthetic, elastomeric adhesive is required, SBR must be considered. Generally, SBR is not used to bond metal parts because of the inferior tack and film strength compared to natural rubber. However, solvent-based adhesives are used to bond plastics, rubber, wood, and fabrics.

Phenolics. As previously discussed under chemically reactive adhesives, phenolic resins represent the largest volume commodity among the synthetic adhesives. For many applications, phenolic-based glues are deposited on a substrate from aqueous, alkaline solution or the resin is handled on a 100 percent solids basis as an uncured syrup. However, for wood bonding (plywood) dilution of the phenol- or resorcinol-derived adhesive with an organic solvent is advantageous for handling, mixing, and coating. In addition to plywood manufacture, other uses of solvent solutions of phenolic resins are in the preparation of wood particle boards, the bonding of glass-fiber mats, and in aircraft honeycomb construction. Blends with synthetic rubbers have already been described. High strength compositions can be prepared by modification with thermoplastic resins.

Urethanes. While urethane adhesives are normally formulated so that a cured or thermoset product can be obtained, as already described under plural component reactive systems, the polyisocyanates are usually diluted with dry solvents such as methylene chloride or chlorobenzene for easy application. Mixtures with rubber modifiers are also prepared in this manner. The reactive isocyanate can be coated on the substrate and the solvent allowed to evaporate. The cure process begins after exposure to atmospheric moisture. This technique has been used with success for bonding rubber to fibers, rubber to metals, and for wood assemblies. For these applications, solutions of triphenylmethane triisocyanate (TTI) or diphenylmethane diisocyanate (MDI) are generally used.

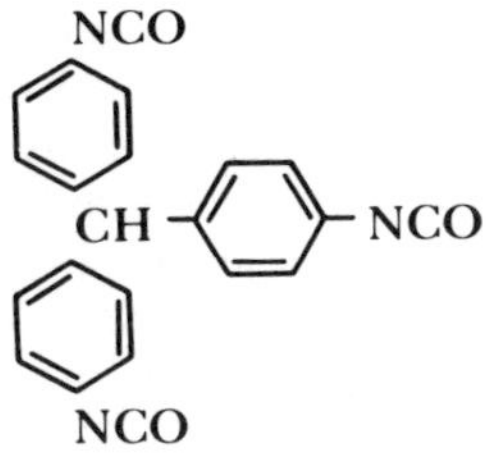

TTI

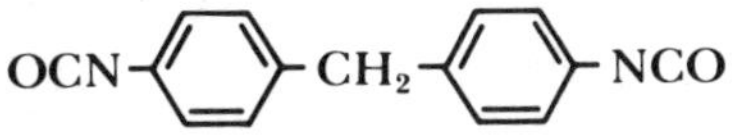

MDI

Two-part urethane formulations in solution are used to laminate packaging films. Polyester can be bonded to polyethylene films with this system. Thermoplastic urethanes are also available for use in solvent-based adhesives.

Vinyl Resins. Several vinyl monomers are used to prepare thermoplastics which are useful in certain adhesive applications. Thermoplastic resins do not "cure" when heated but develop joint strength by solvent evaporation, diffusion, or by fusion. The most important vinyl resins for adhesives are polyvinyl acetate, the polyvinyl acetals (butyral and formal), and polyvinyl alkyl ethers. Polyvinyl chloride and copolymers of both vinyl chloride and vinyl acetate (general formulas below) with other monomers such as maleic acid esters, alkyl acrylates, maleic anhydride, and ethylene are also used for solvent-based adhesives.

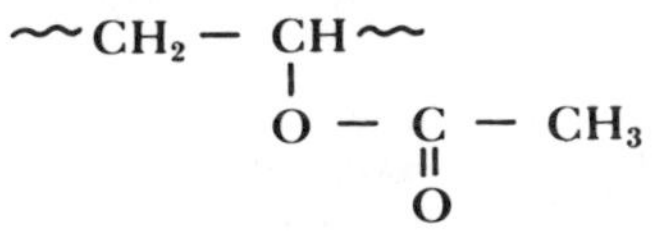

Polyvinyl acetate

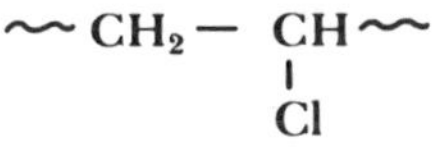

Polyvinyl chloride

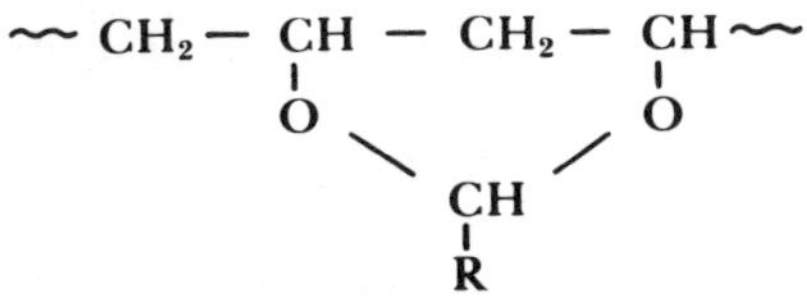

Polyvinyl acetal

Polyvinyl acetate resins can be dissolved in alcohols, ketones, and esters for coating on substrates. Clear films deposited from solvents can be heat sealed to substrates for satisfactory adhesion. Metal foil to plastic film or paper laminates can be manufactured in this manner. Polyvinyl acetate and its copolymers are also used extensively in hot melt adhesive compositions. Copolymers with vinyl chloride are used for bonding vinyl plastic film to metals.

The polyvinyl acetal resins represent an important part of the solvent-based vinyl adhesives category. Polyvinyl butyral is employed as the basic resin component for the laminating adhesives in the manufacture of safety glass. Acetal-phenolic resin mixtures are used extensively as structural adhesives for metal-honeycomb systems and metal to wood laminates.

Other vinyl resins which are occasionally used as solvent based adhesives are polyvinyl alkyl ethers and polystyrene. Solution of the vinyl ethers are used where a high degree of tack is needed, and polystyrene is used for bonding polystyrene to itself and to porous materials such as wood and plaster.

Acrylics. While not as important as the vinyls in terms of total usage in the adhesives industry, a number of acrylic resins are used for bonding cloth, plastics, leather and in some cases metal foils. The acrylic monomers most commonly used in adhesives are:

$$CH_2{=}CH-\overset{\overset{\displaystyle O}{\|}}{C}-O-CH_2CH_3$$

Ethyl acrylate

$$CH_2{=}\overset{\overset{\displaystyle CH_3}{\backslash}}{C}-\overset{\overset{\displaystyle O}{\|}}{C}-O-CH_3$$

Methyl methacrylate

$$CH_2{=}\overset{\overset{\displaystyle CH_3}{|}}{C}-\overset{\overset{\displaystyle O}{\|}}{C}-OH$$

Methacrylic acid

$$CH_2{=}CH-\overset{\overset{\displaystyle O}{\|}}{C}-OH$$

Acrylic acid

$$CH_2{=}CH-\overset{\overset{\displaystyle O}{\|}}{C}-NH_2$$

Acrylamide

$$CH_2{=}CH-CN$$

Acrylonitrile

The polymers or copolymers are soluble in common organic solvents such as chloroform or toluene and can be applied in the same manner as other solvent-based adhesives. In addition, the polymers are soluble in the monomers, and when a catalyst is added, the solvent monomer will polymerize to give good adhesive strength to glass and to plastic surfaces of similar composition (e.g., polymethyl methacrylate).

The advantages of the acrylics over the vinyls for adhesive use are:

1) They are stable in sunlight
2) Some are flexible at low temperature
3) Some of the acrylic copolymers are oil resistant
4) They are readily modified with cross-linkable monomers so that heat cured adhesives with improved solvent resistance can be obtained.

Such systems are useful in textile laminating applications where resistance to dry cleaning fluids is important. The acrylic resins are generally more expensive than vinyl resins.

One class of acrylic adhesives differs from those already discussed in that solvents are usually not employed in handling; these are the cyanoacrylates (38) (general formula below). These materials are activated by pressing into thin

$$CH_2 = \overset{\overset{\displaystyle CN}{|}}{C} - \overset{\overset{\displaystyle O}{\|}}{C} - O - CH_3$$

Methyl 2-cyanoacrylate

films between the substrates. Very rapid curing (10 to 30 seconds) occurs, and good bonds develop with glass, metal, rubber, and wood. The chief disadvantage of this material is its high cost. The major use is in bonding small parts where a quick set is required. The electronics and jewelry industries utilize cyanoacrylates in operations where it is difficult to clamp or hold the assembly. Ethylene glycol dimethacrylate is also used for this kind of quick application.

Miscellaneous Resins. A few materials are used in a more limited way as solvent adhesives. Two natural product-derived materials are the cellulose esters and asphalt. Cellulose esters are used for bonding leather, paper, and wood. While not generally used with metals, specific non-porous substrates such as cellophane and glass are occasionally bonded with cellulose nitrate or other cellulose esters applied from solution. Asphaltic materials from petroleum fractionation are inexpensive products which are used for some special adhesive applications. The most important property of asphalt in adhesives is water resistance. It is used in laminating paper and other packing materials where a water barrier layer is needed. The other important use of asphalt is in roofing and flooring adhesives.

Another class of resins which is used in solvent adhesives is the polyamides. The high molecular weight products are generally referred to as modified nylons. Low and intermediate molecular weight materials are also available. The latter two are more commonly used in hot melt formulations, and the modified nylons methyl methoxy (modified terpolymer) are often blended with small amounts of a phenolic resin to improve surface wetting. Solution systems of the low and intermediate molecular weight resins can be coated on paper, metal foil, or plastics and when heat activated will act as adhesives for these substrates. The modified nylon polyamides have fair adhesion to metals and have good low and high temperature properties, good resistance to oils and greases but have poor resistance to solvents.

Several recently developed thermoplastics are beginning to be used as solution-applied adhesives for selected applications. Examples are phenoxy resin, bisphenol A polycarbonate, and polysulfone. These materials are temperature resistant plastics which require fairly high temperatures for heat activation after solvents have been removed. Phenoxy has been used quite extensively as an adhesive and will develop lap shear values of 3500 to 4000 psi in a few seconds at 600°F surface temperatures and in 2 to 3 min at 500°F. Good adhesion is observed with substrates such as copper, brass, steel, aluminum, wood, and many other nonmetallic substances.

Water-Based Systems

Water solutions, dispersions, or emulsions are widely employed for application and bonding of various substrates. Water systems are most frequently used when at least one surface to be bonded is porous, so that the water may escape by permeation. The major advantage of water as a vehicle for adhesives is the lower cost compared to organic solvents. Other advantages of water-based adhesives are nonflammability, low toxicity, and the wide range of viscosity and content of solids which can be obtained. Disadvantages are generally poor water resistance, slow drying, a tendency to freeze, and a shrinkage of fabric substrates.

Natural Rubber. As mentioned under solvent systems, natural rubber is isolated as a water dispersion known as latex. Thus, the material can be used directly as a water-based adhesive. Usually the latex is treated with preservatives such as ammonia and formulated for specific end uses. Natural rubber latex dries to an almost tack-free film which will, however, adhere to itself. This property is utilized in the manufacture of dry-seal envelopes. Generally, natural latex is blended with other adhesive raw materials such as phenolics, casein or natural tackifiers like Kayara gum to obtain better bonding. Portland cement blends can be used to give adhesion to metals, glass, and cementitious materials.

Reclaimed Rubber. Reclaimed rubber is used in a water dispersed formulation for a number of adhesive applications. Mixed with natural or synthetic rubber latexes, reclaimed rubber provides a lower cost adhesive system. Asphalt blends have excellent tack and are used to join textiles to metals, felt to metal, and similar applications where an inexpensive adhesive in a nonflammable solvent is required.

Synthetic Rubbers. All of the synthetic rubbers discussed under solvent adhesives are available as emulsions in water. Generally, butyl rubber is used in solvent solution, but nitrile, neoprene, and styrene-butadiene latexes represent an important part of the market for synthetic rubber adhesives. Styrene-butadiene rubber latex is used for carpet backing, paper coating, and tire cord bonding. Neoprene latex is used in paper treatments to give leather-like properties, in casein blends to laminate metal foil to paper, and in shoe adhesives. Nitrile rubber latex is used as a binder for nonwoven fabrics and in paper treatments.

Vinyl Resins. By far the most widely used vinyl resin in water dispersion form is polyvinyl acetate both as homopolymer and copolymer. Prepared by an emulsion polymerization technique, polyvinyl acetate latex is the basis for the common household "white glue." Products of this type are good adhesives for paper, wood, leather, and cloth. The major end use is as a packaging adhesive for such flexible substrates. Polyvinyl acetate water dispersion is also used as a lagging adhesive to bond insulating fabric to pipe and ductwork in steam plants and ships. When suitably formulated, satisfactory adhesion can be developed to plastic surfaces and to metal foil.

Among the other vinyl resins, polyvinyl alcohol, and polyvinyl methyl ether are the most useful for water-based adhesives. Polyvinyl alcohol is prepared by hydrolyzing polyvinyl acetate, and when the degree of hydrolysis is from 88 to 99 percent, the resin is quite soluble in water. The water solutions are used to glue paper and cloth, particularly in food packaging applications where lack of toxicity and odor are definite advantages. Polyvinyl methyl ether is also water soluble and is used to increase the tack of water-based systems and to increase the viscosity of aqueous solutions of adhesives. Since polyvinyl methyl ether is water soluble, it can be used as a remoistenable adhesive.

Acrylics. A variety of acrylic copolymers are prepared commercially by emulsion polymerization. The most common monomers are the ones listed in the solvent adhesive section under acrylics. The isolated latex emulsions are frequently used as components of water-based adhesives. Crosslinking monomers are sometimes built into the latex formula, so that when heated, a solvent resistant bond is developed for dry cleaning resistance in fabric laminates. Advantages and disadvantages of the acrylic latex products are similar to the solvent adhesive properties outlined earlier.

Miscellaneous Resins. Some other polymeric materials which are used in water solution or dispersion form as adhesives should be mentioned. Water-sensitive materials derived from natural products include animal glue, starch, soya and blood glue, casein, and certain cellulose derivatives. Of these, starch is used in the largest volume primarily for paper packaging and as a binder for clay coatings for paper.

When carboxyl groups are present at sufficient concentrations in synthetic polymers, the resins become soluble in aqueous alkali. Examples of such polymers are ethylene-acrylic acid and ethylene-methacrylic acid copolymers, vinyl acetate copolymers containing maleic acid or crotonic acid, and acrylates modified with unsaturated carboxylic acids. These carboxyl-containing copolymers exhibit enhanced metal adhesion and grease resistance in comparison with the same systems without the acid groups.

HOT MELT TYPES (COOLING)

Several major industries have recently converted to the new hot method of applying adhesives. This method offers the advantages over older techniques of high speed production rates and simplicity in handling the adhesive itself. The packaging, bookbinding, and shoe industries are most involved in hot melts, using up to 40 million pounds of adhesive per year, with further rapid growth anticipated.

The cover of this book is a good example of hot melt bonding of the first type that will be described. A visit to the grocery store will also reveal many new ways of packaging—six-pack containers, folding cartons, sealed see-through wrappings for bacon, razor blades, etc.—all using hot melt adhesives (see also Fig. 1–8) (39).

A hot melt may be defined as a 100 percent solids thermoplastic adhesive polymer. The definition would seem to include *all* thermoplastics, which is potentially true. However, considerations in applying the adhesive and forming the bond will limit the hot melts to roughly ten types. These materials are ideally solid up to a temperature of 175°F or higher, then melt sharply to give a low viscosity fluid that is easily applied and capable of wetting the substrate to be bonded, followed by rapid setting (matter of seconds) upon cooling. When hot melts are used, a new set of polymer properties such as: softening point, melt viscosity, melt index, crystallinity, tack, heat capacity, and heat stability must be understood and used in addition to the usual physical and strength properties. Chapter 7 will provide more details in this area.

The adhesives themselves are not necessarily newly developed materials. It is the combination of *properly formulated resins* and of *application equipment* to handle these resins that has made the hot melt technique successful (40). This chapter discusses only the adhesive characteristics. Table I–11 gives representa-

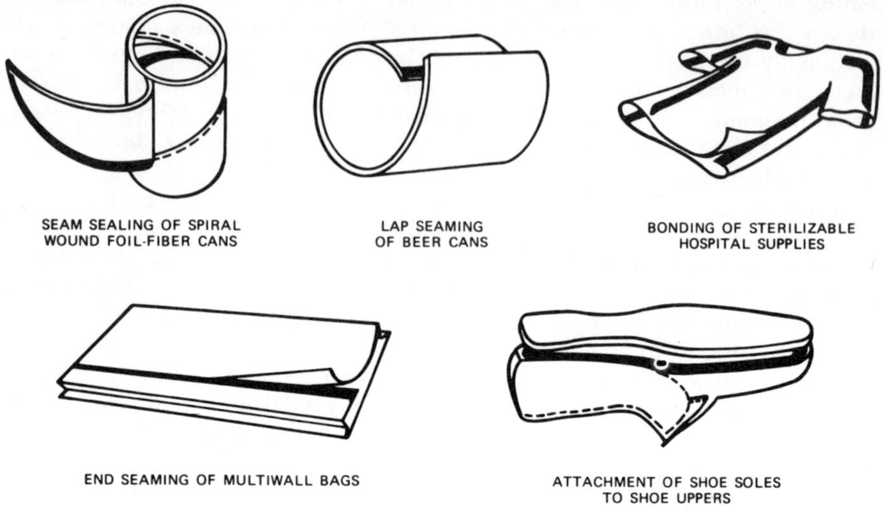

Fig. 1–8. Examples of hot melt technology (39).

tive data as a general comparison between adhesive types. An individual hot melt is usually available in a wide range of melt and flow properties, and the formulated hot melt available to the user will contain several major components (see Fig. 1–9).

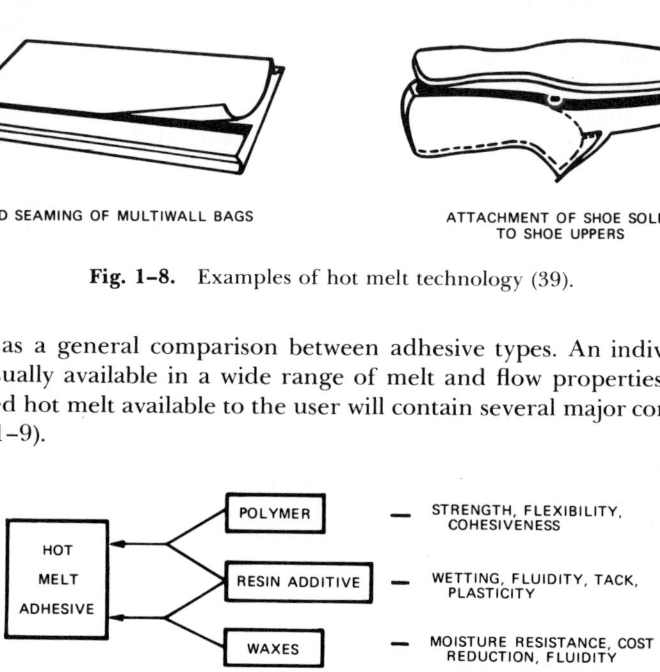

Fig. 1–9. General comparison of hot melt formulated adhesive properties.

Adhesive Types—General

Polyethylene is the leading synthetic polymer in production today. Its low cost makes it an attractive candidate for hot melts. However, a deficiency in good adhesion requires it to be modified with other chemical groups that give stronger bonds. Thus a series of ethylene copolymers has been introduced that includes (41):

1) E/VA (ethylene vinyl acetate)
2) E/EA (ethylene ethyl acrylate)
3) E/IBA (ethylene isobutyl acrylate)
4) E/AA (ethylene acrylic acid)
5) "Ionomers" (ethylene acrylic acid salts)

Table I–11. General Comparison of Hot Melt Adhesives

	E/VA	*E/EA*	*E/AA*	*Iono-mer*	*Phe-noxy*	*Poly-amide*	*Poly-ester*	*Poly-ethyl-ene*	*Poly-vinyl Ace-tate*	*Poly-vinyl Bu-tyral*
Softening Point, °C	40	60	70	75	100	100	—	—	65 to 195	—
Melting Point, °C	95	90 (80 to 100)	—	—	—	—	267	137	—	130
Crystallinity	L	L	M	(L)	L	L	H	H or L	L	L
Melt Index	6	3 (2 to 20)	3 (0.5 to 400)	2	2.5	2	5	5 (0.5 to 20)	—	—
Tensile Strength, psi	2750	2000	3300	4000	9500	2000	4500	2000	5000	6500
Percent Elongation	800	700	600	450	75	300	500	150	10	100
Lap Shear on Metal, psi	—	—	1700	—	3500	1050	—	—	—	—
Film Peel, lb/in	1	4.5	12	12	—	—	5	2	—	—
Cost	M	M	M	M	H	M	H	L	M-L	M
Usage	H	M	L	L	L	H	H	L	M	H

H = High
M = Medium
L = Low

Other major groups of hot melts include: (1) polyvinyl acetate, widely used as a base material or additive to other polymers; (2) cellulose ester derivatives, used where elevated temperature performance is needed, and (3) polyamides and polyesters, used where strong adhesion is required at a higher allowable cost. Individual types of hot melts are discussed briefly in the following section.

Olefin Copolymers

Ethylene Vinyl Acetate (E/VA). The polyethylene structure (general formula below) is occasionally broken by the appearance of the highly polar and adhesionable acetate group.

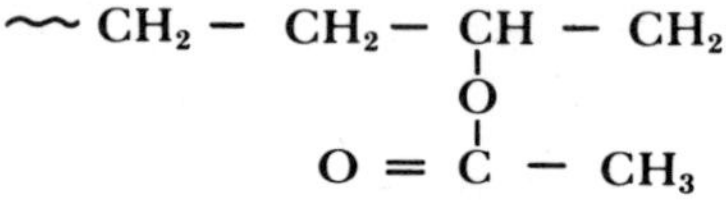

Mol Wt ≈ 5000

The E/VA series of hot melts has acquired a leading role in the bookbinding and packaging fields within the past three years. It is also used in "instant heat-seal" labels and affords: (1) compatibility with the other necessary components, (2) a range of products varying in melting points, and (3) relatively low costs. The melt point will average 167°F (75°C) minimum to 203°F (95°C) average. Table I–12 lists examples of bonding with E/VA and related hot melts for several substrates. Adhesion values are given in grams per inch width.

Table I–12. Peel Values of Ethylene-Based Adhesives.

Bonded Material	*E/VA*	*E/EA*	*E/IBA*	*PE*	*Wax*
	(Adhesion Values in gms/in width)				
Paper to Polyethylene	320	390	390	—	—
Paper to Kraft	400	320	300	330	50
Paper to Aluminum Foil	350	130	250	110	10
Paper to Polyester Film	20	140	—	2	15

Legend: E/VA = Ethylene vinyl acetate
E/EA = Ethylene ethyl acrylate
E/IBA = Ethylene isobutyl acrylate
PE = Polyethylene

Ethylene Ethyl Acrylate (E/EA). Although different from E/VA (above), E/EA (general formula below) is very similar. The use of ethyl acrylate in place

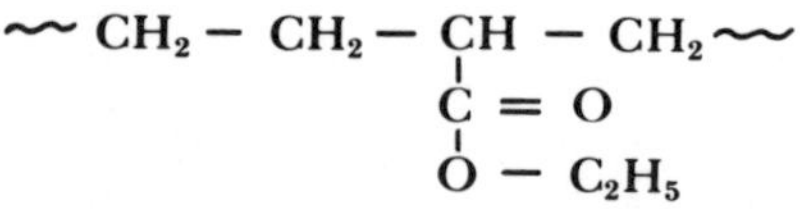

of vinyl acetate gives better heat stability, toughness, improved adhesion to a wider range of substrates, and better compatibility with the other components. However, it has been demonstrated that often *marginal* compatibility gives better performance than complete solubility of all components.

Ethylene isobutyl acrylate (E/IBA) is an alternate product and may be recommended by a manufacturer over E/EA for specific application (adhesion to polypropylene). These types, when combined with E/AA or E/VA, produce adhesives that will bond together two widely differing substrates.

Ionomers (42). Ionomers (general formula below) can be regarded as a

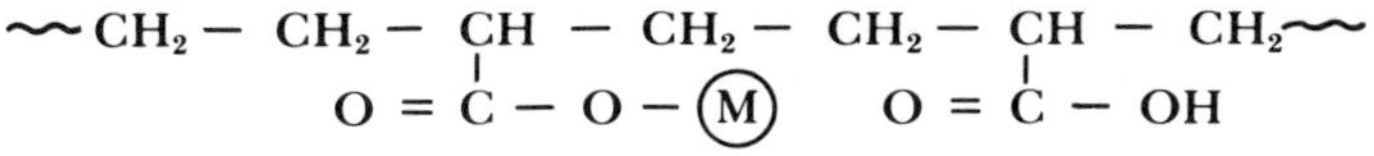

(M) is a metal cation

derivative of the ethyleneacrylic acid (E/AA) copolymers. It is available both in film form and as a hot melt. Ionomers display good adhesion to most metals and nonmetals, but the bonds are somewhat sensitive to moisture. Since 180°F is the softening point, this is the maximum temperature for most uses, although good peel strength is retained at 200°F or higher.

Ionomers differ from the E/AA polymers in that the ionomers have a metal cation present on some of the pendant carboxyl groups. The presence of both cations and anionic side groups allows interchain bonds to form between adjacent molecules. These bonds can be considered as analogous to the cross-linking process in a thermoset resin, but here the bonding is thermally reversible. This results in improved tensile strength: 4000 psi vs 2750 psi for a corresponding E/VA or 3350 psi for an E/AA copolymer.

Phenoxies

Phenoxies (general formula below) are hot melt adhesives with unusually

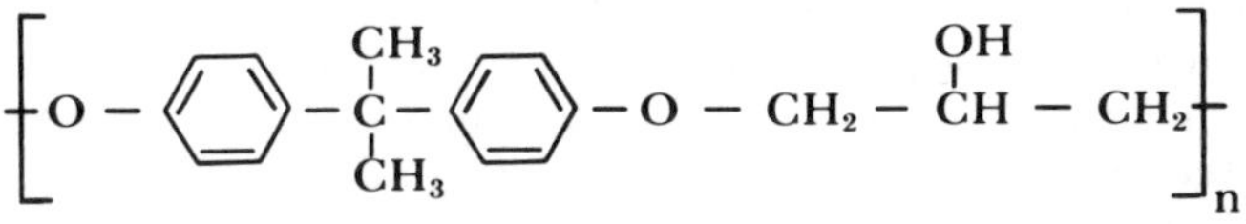

high strength, which is advantageous when bonding to metal substrates. The high application temperature of from 500°F to 600°F also makes these adhesives natural for bonding to metal substrates.

The chemical structure is similar to that of the typical epoxy resins. Thus the tensile strength is high at 9000 psi. Lap shear strength will be 3500 to 4000 psi tested at room temperature or 2800 psi at 180°F. This is in the range for the usual structural adhesives for metals.

Phenoxies may be modified with other resins such as polyvinyl ether, to reduce the bonding temperature requirement. High cost and the high melting point restrict the use of phenoxies.

Polyamides

The polyamides (general formula below) fit into three separate adhesive

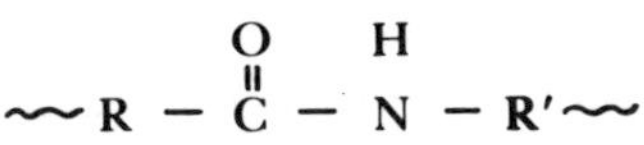

categories (43):

1) Low molecular weight—liquid or solid. Used as a thermoset (with epoxies) and also as a hot melt adhesive.
2) Intermediate molecular weight—solid. Used as a hot melt adhesive.
3) High molecular weight—solid (nylons). Used in thermoset films (with epoxies) and hot melt adhesives.

Polyamides have the advantage of combining good strength with flexibility. Hot melts find their major usage in shoe bonding. They are also used as can-seam sealants and to a lesser degree in producing laminated packaging film. The low and intermediate molecular weight (mol wt) polyamides are based on unsaturated dibasic acids of vegetable origin. These acids are "dimerized" and then reacted with diamines, as shown in the formula below:

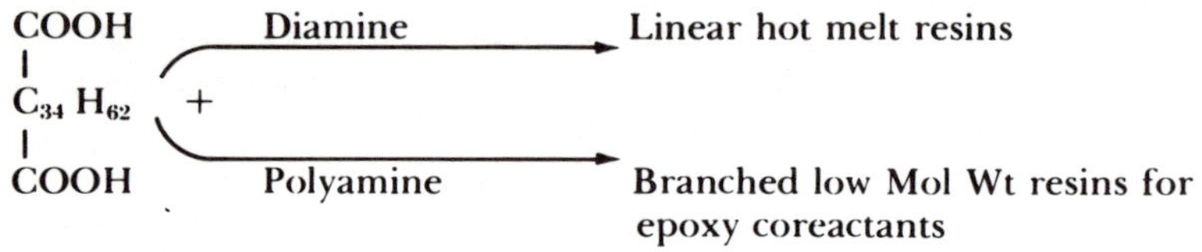

As hot melts (44), the dimer acid-based polyamides have these features:

1) Range of melting points
2) Range of melt viscosities
3) Good adhesion to a variety of substrates
4) Fast set
5) Good solubility, particularly in the lower mol wt versions.

The melted resins are unstable in air above 440°F. Hot resin, therefore, must be protected by a nitrogen blanket, as well as by stabilizers and anti-oxidants.

Polyamide hot melts can be formulated with phenolics, rosins, waxes, and plasticizer such as sulfonamides. The adhesives are used as can-seam sealants. Another major use is in shoe fabrication. Some "difficult-to-bond" substrates for which the newer polyamides will show good adhesion are neoprene, ABS plastic, and plasticized vinyls.

Recently, a series of high molecular weight polyamides (nylons) have been introduced which are chemically related to these dimer acid-based polyamides. The characteristics include the excellent substrate wetting properties of the dimer-based polyamides and the high strength and toughness of the nylons. Application with conventional hot melt equipment is difficult because of their high melt viscosity; however, highly successful applications have been developed to extrude these nylon resins onto metallic substrates.

Polyesters

Saturated polyesters as hot melts are replacing neoprene adhesives in the shoe industry. The advantage of polyester is high crystallinity which gives first a sharp melting point (a decided advantage in high speed bonding) and often gives a combination of high tensile strength plus a high percentage of elongation in a polymer. Molecular weight should be adjusted for good melt fluidity. The value should be low but at a minimum of 4000 to 5000 for suitable tensile strength.

Tensile strength runs from 4000 to 5000 psi, and film peel strength runs from 4 to 6 lb/in when bonding plastic to plastic. As was noted in the section on Chemically Reactive Types, saturated polyesters adhere best to polyester substrates, promoting some usage in the lamination of packaging films (more often as two-part systems in solution than as hot melts). These may use a special hardener or catalyst containing isocyanate to promote adhesion.

In the shoe industry polyesters are used in three areas: (1) attachment of shoe soles, an exacting high speed operation, (2) lasting (shaping of the upper part) which is less demanding in assembly techniques, and (3) folding.

Polyethylene and Polypropylene

Polyethylene as an adhesive (general formula below) is limited by its very

$$\sim CH_2 - CH_2 - CH_2 - CH_2 \sim$$

or

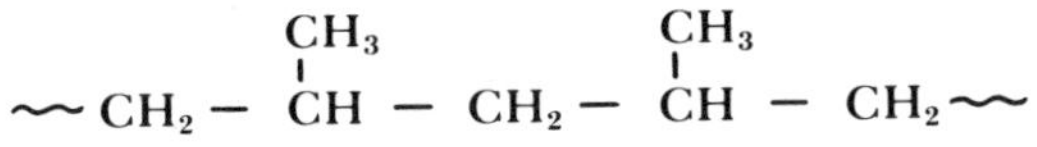

poor adhesion, but is desirable because of its low cost. It is used widely as a heat seal coating on paper and plastics. Linear high density PE is partially crystalline which gives better strength and a sharper melting point at 220 to 260°F. The branched low density polymer is easier to work as a hot melt due to better melt fluidity. Polyethylene with a wax modifier is used in paperboard bonding and in paper to paper spot heat seals such as on milk cartons, etc. Polyethylene also has found limited use as a flocking adhesive for automotive carpeting. The non-crystalline form of polypropylene is by nature a tacky, rubbery but high melting material. It is used in blends with other hot melts in the packaging field.

Polyvinyl Acetate (PVA)

Along with the cellulosics, polyvinyl acetate (PVA) (general formula below)

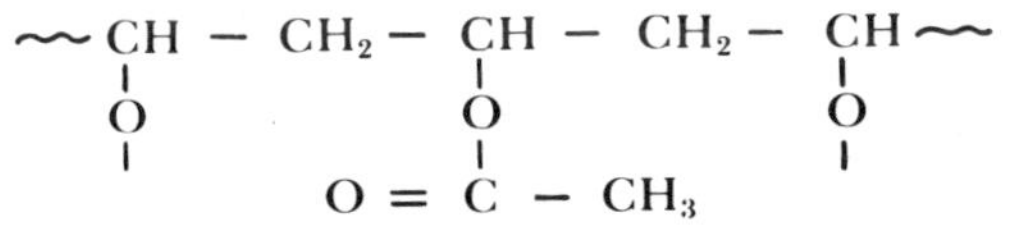

is one of the first polymers to be used as a hot melt. Newer resins have superseded these for many applications, but PVA still has wide usage in the frozen food packaging area where flexibility at low temperatures is important. Other uses include bookbinding and the lamination of foils.

Polyvinyl Butyral

When used as a hot melt, polyvinyl butyral (general formula below) gives

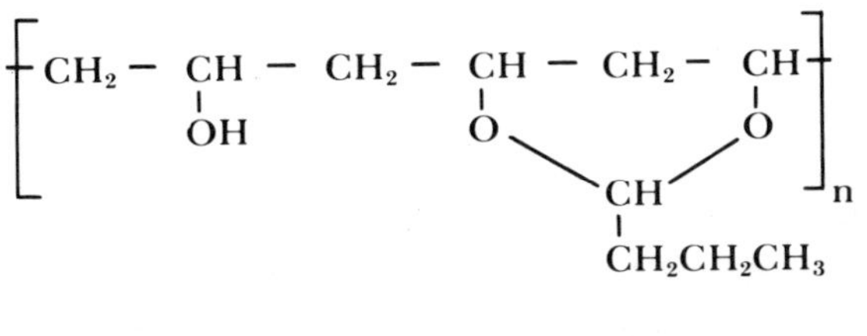

a tough, clear glue line. The melting point is from 257°F to 302°F (125° to 150°C). It is formulated with a plasticizer such as diphenyl phthalate, waxes, and rosins. Data on adhesion to glass with a polyvinyl ether additive is given in the following section.

Polyvinyl Ethers

Water solubility (45) is the unusual property of polyvinyl ethers (general formula below). This allows it to be used in remoistenable adhesives (gummed

$$\sim CH_2 - \underset{OCH_3}{\underset{|}{CH}} - CH_2 - \underset{OCH_3}{\underset{|}{CH}} \sim$$

labels) and as an additive to other systems to promote bonding to moist surfaces. A prime example is its use with polyvinyl butyral in safety glass lamination. High humidity conditions enhance its adhesion to glass. Its uses as a pressure-sensitive adhesive are discussed later in this chapter.

Polyvinyl ethers find some use as hot melts in bonding labels to metal surfaces and in bookbinding. The cohesive strength is low, encouraging usage primarily as an additive to other systems. A 1:1 blend of polyvinyl methyl ether with a polyamide, formulated with rosin, wax, etc., is an example of a use of polymer to improve adhesion to plastics and metal foils.

The relative bonding effectiveness of polyvinyl butyral blended with different percentages of the additive polyvinyl methyl ether (PVME) on various moisture-laden surfaces is shown in Table I–13. Moisture resistance of polyvinyl butyral is poor until PVME is added; the percent of adhesion increases with the increase of PVME.

Table I–13. Effectiveness of Automotive Safety Glass Adhesives Using PVME* Additive to Promote Moisture Resistance (45).

	Percent of Adhesion	
Adhesive	*"Low Moisture" Surface*	*"High Moisture" Surface*
Polyvinyl Butyral	80	10
Polyvinyl Butyral, 5 percent PVME*	90	60
Polyvinyl Butyral, 10 percent PVME*	95	80

*PVME = polyvinyl methyl ether.

Asphalt

Asphalts (general formula below) are used by the millions of tons in paving

$$\sim \begin{matrix} \text{Mixed} \\ \text{Hydrocarbons} \\ \text{Mol Wt} \approx 2000 \end{matrix} \sim$$

and roofing. These are hot melt coating applications, and these greatly overshadow the adhesive uses. The latter area includes hot melts, mastics for flooring, and adhesion extenders (46).

Asphalts are natural products; the principal sources are native deposits or residue from petroleum refining. Cost is very low, from 1 to 2 cents/lb. The ability of asphalt to serve as a moisture barrier is its main advantage. Otherwise, despite the disadvantages of low tensile strength and poor color, asphalt compares with the synthetic plastics in shock and creep resistance. Uses include high speed production of hot melt laminated kraft paper in which the asphalt serves as a moisture seal on paper covered building panels and insulation boards, and as an additive to latexes to reduce cost and improve moisture resistance necessary for the adhesion of metal to fabrics.

Ceramics

Ceramics (general formula below) can be considered as a special class of hot

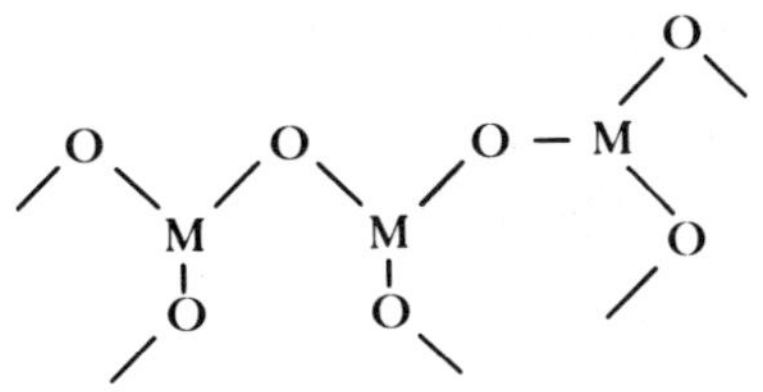

M = Al, Si, Ca, B, Mg, etc.

melts; they are the only materials available for service at 1000°F to 1500°F. Thus they are used in special aerospace applications in spite of brittleness and very low peel strength. The adhesives are supplied either as a water slurry or as a coated metal wafer. In either case, the bond is assembled and then fired at perhaps 1700°F to fuse the ceramic. For service at 2000°F, the bonds must be "matured" at 3200°F (47).

Cermets, a combination of the ceramic with a metal or brazing alloy (e.g., lead borosilicate glass plus metallic cobalt), alleviate the brittleness to a degree (48). Uses include metal-to-metal bonding and honeycomb construction.

As an alternate to the fusion technique, room temperature setting cements are available. Compositions based on aluminum phosphates and silicates are available. These require mixing of two components or the addition of water. When properly cured, lap shear strength will reach 500 psi on metal at room temperature, or 300 psi at 2000°F with ceramic substrates.

DELAYED-TACK ADHESIVES

A delayed-tack adhesive differs from both a hot melt adhesive and a heat-seal adhesive. A hot melt will be non-tacky when cooled, while a heat-seal adhesive will seal under heat and pressure but does not remain tacky for any period of time after cooling. On the other hand, after a delayed-tack system is heat activated and cooled, it will remain tacky anywhere from several minutes to several days over a wide temperature range. For example, a rubber-based delayed-tack adhesive plasticized with diphenyl phthalate can become heat activated at 150°F. The formulation remains adhesive and tacky even as the temperature cools to 50°F. Below 50°F, the adhesive will set.

Delayed-tack adhesives are coated on paper for labels on such items as bread packages and càns. To obtain the delayed-tack characteristics of the adhesive system, a solid plasticizer must be incorporated. A solid plasticizer also contributes to non-tackiness before heat activation. The most commonly used solid plasticizers are dicyclohexyl phthalate, diphenyl phthalate, N-cyclohexyl-p-toluene-sulfonamide and o/p-toluenesulfonamide. Because of the range of melting points available, adhesives with different heat activation temperatures are possible.

The most commonly used resins are styrene-butadiene copolymers, polyvinyl acetate, polystyrene and polyamides.

FILM ADHESIVES

Bonds produced by a given adhesive will usually have similar properties whether the adhesive is applied as a one-component paste or as a solid film. The advantages of using the adhesive in film form are (49):

1) Uniform, controlled glue line thickness
2) Speed and ease of application (a clean, solvent-free operation is allowed)
3) Two-sided films can be prepared. These are used in lightweight sandwich construction. The honeycomb side of the adhesive will provide good filleting; the skin side will provide high peel strength. If one side of the film is tacky, it is easier to align the assembly to be bonded.

In some film adhesives, a woven or knitted fabric that may be used to support the polymer film will also carry part of the load and will give improved bond strength by better distribution of the applied forces.

Film adhesives are provided in two forms; unsupported, or alternately supported on a flexible carrier such as glass cloth, nylon, or paper. The carrier will usually have little effect on adhesive properties.

Epoxies

Epoxy alloys have surpassed the original structural film adhesives, nitrile-phenolic alloys, first devised for aircraft assembly. Usage of the epoxies has spread to other areas such as electrical systems fabrication and building panels.

The epoxy resin is usually a blend of the bisphenol A type plus an epoxy novolac. The latter is a solid resin with a high concentration of reactive groups and good cured properties at elevated temperature. The customary curing agent is dicyandiamide. This is added to the epoxy as a solid powder that is insoluble in and unreactive with the epoxy until it approaches its melting point of approximately 400°F. A tertiary amine may be added with controlled preheating to form a "B-Staged" resin in which the DICY is solubilized and thus evenly distributed in the resin system. Recent developments include use of accelerators to reduce cure temperature from 360°F to 250°F, with 180°F as an objective for a one hour cure.

Peel strength is necessary in a structural epoxy unless the adhesive is applied to a joint so designed that no peel forces occur. Methods of developing peel strength were already discussed in connection with phenolics. Essentially a thermoset resin is combined with a thermoplastic to impart toughness and flexibility to the high strength thermoset. The end product may be called an "engineering plastic," having a good balance of *all* important physical properties.

Examples of alloying polymers for epoxy include nylon, urethane, acrylics, vinyls, and nitriles. With the more elastomeric polymers, the property of resiliency is emphasized.

High peel strength can alternately be provided in an epoxy film adhesive through special techniques in preparing the B-staged film.

Most of the epoxy film adhesives are utilized in aircraft assembly—fairings and lightweight honeycomb. Phenoxies, related to epoxies, have been used in film form to assemble automobile bumpers. Data on film performance are presented in Table I–3.

Nitriles

Nitrile is a shortened term for butadiene-acrylonitrile rubber, a type of rubber adhesive resistant to moisture, oils, and solvents. Plasticized polyvinyl chloride may be bonded to aluminum using a nitrile for example, since migration of the plasticizer will not affect the bond. Certain polyesters, urethanes, and, to a lesser extent, polyamides are also among the very few adhesives suitable for this purpose.

In film form, nitriles are alloyed with other polymers such as epoxy, phenolic, or vinyl.

Epoxy. Epoxies are compatible in all proportions with nitrile. Properties shift smoothly from the high strength, rigid epoxy to the flexible, low shear high nitrile end of the range.

Phenolic. A 40/60 nitrile/phenolic will have a lap shear strength of 4000 to 5000 psi on aluminum or stainless steel. This type was the first successful structural adhesive, and a film form was developed early in its usage. Uses include:

1) Metal lamination (including aircraft)
2) Brake lining to brake shoes
3) Panel fabrication (curtain wall, etc.)
4) Leather and shoe bonding.

Vinyl. Nitrile-polyvinyl chloride alloys are useful for hot melt bonding. Adhesion and flexibility are good, and the melt temperature is in the correct range. Films may be used in bonding leather or vinyl substrates.

Olefin Copolymers

Ethylene/Acrylic Acid (E/AA). The free acid groups in ethylene/acrylic acid (general formula below) adhere exceptionally well to metals. Beyond this

$$\begin{array}{c} \mathrm{R} \\ | \\ \sim\mathrm{CH_2}-\mathrm{CH_2}-\mathrm{C}-\mathrm{CH_2}-\mathrm{CH_2}\sim \\ | \\ \mathrm{O}=\mathrm{C}-\mathrm{OH} \end{array}$$

property, the material closely resembles its corresponding unmodified polyethylene analogue. E/AA copolymer is most often formed into a film adhesive, but it also is used as a hot melt.

In addition to metals, this adhesive will bond polyethylene, cellulosics, nylon, urethanes, epoxies, etc. (50). Metal bonding uses a 15 percent AA copolymer;

nonmetals use 3 percent AA. Properties of the bonds prove best when measured dry at room temperature. For example, peel strength to aluminum is 60 lb/in at room temperature but drops to 10 lb/in in measuring either at 32°F or at 150°F. The initial value drops to 43 lb/in after a 24-hr water soak. A lap shear strength of 2000 psi has been measured on aluminum substrates.

Fig. 1–10 shows the importance of a high bonding temperature in promoting peel strength to a metal substrate. This curve is characteristic of film adhesives in general. A very similar curve is also shown for the ionomer film discussed in the next section. Some theoretical attention is being paid to data of this nature.

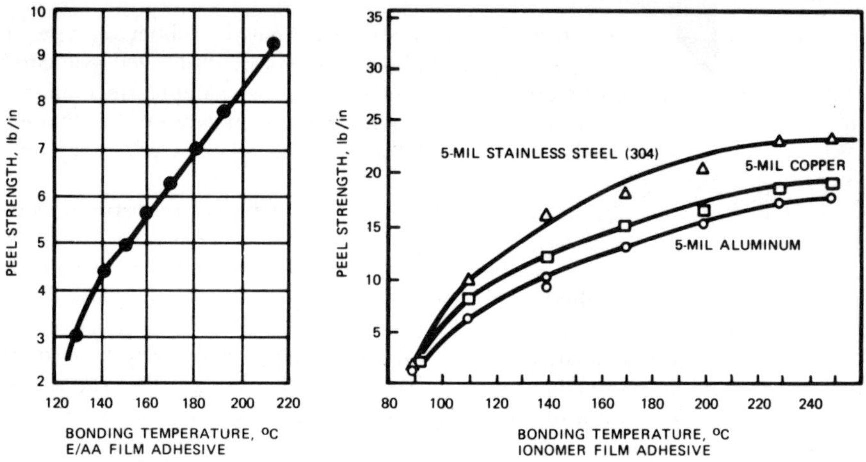

Fig. 1–10. Effect of bonding temperature on the performance of film adhesives: left, E/AA film adhesive (film to metal bonding) *(Courtesy, The Dow Chemical Company)*; right, ionomer film adhesive (metal to metal bonding) (51).

Ionomers. Ionomers are easier to handle as films than as hot melts. This is often true for adhesives available in either form.

Temperature of bond formation (51) is important in determining strength and moisture resistance of the cured bond (Fig. 1–10). The higher the cure temperature, up to the point of thermal decomposition of the polymer, the better the peel strength and the shorter the cure time requirement. Full strength can be developed in 5 min at a 400°F cure temperature.

Film thickness is important in giving high peel strength. A 3-mil thickness is usually adequate in metal-to-metal bonding (for any adhesive), but a 5-mil ionomer film gives better peel values on thick metals. Temperature range for the cured bonds is similar to the range for the E/AA film—best at room temperature but serviceable from approximately 30° to 200°F. Creep strength is reported to be good.

Ionomer and E/AA films are both used in producing metal laminates (52). Metal/nonmetal composites include plates for printed circuits. Metal/metal bonding includes:

1) Stainless steel/aluminum, for decorative low cost sheet.

2) Copper/steel, to conserve copper metal. The laminate combines electrical conductivity (copper) with a high strength low cost steel backing.
3) Laminated metal. Alternate plies of thin metal and adhesive are compiled to the final sheet or plate thickness. Plate of ½ in. total thickness has been made from $^1/_{16}$ in. aluminum, bonded with a two-part epoxy. The plate has twice the fracture toughness of conventional monolithic plate. Metal-adhesive composites of this type are found to have additional desirable properties in the area of vibration damping, sound deadening, and fatigue resistance.

Phenolics

Phenolic resins supported on a paper carrier are used in specialty plywood products. This adhesive is much more costly than the typical liquid phenolic resin for plywood; therefore savings accruing from processing or from the shortening of assembly time must enter into the decision to justify the use of the film.

Structural adhesive films are prepared by alloying with epoxies, nitriles, vinyls, etc. These are preferred to other film adhesives when high temperature performance is needed. For example, an epoxy-phenolic can withstand short exposures at 1000°F. A conventional epoxy would be limited to 400°F maximum. Otherwise in comparison to epoxies, phenolics have lower strength and require high pressure (100 to 200 psi) during bonding since volatiles are released. If cost is a factor, the vinyl-phenolics should be considered.

Polyamides

The intermediate mol wt polyamides that are used in hot melts (see Hot Melt Types (Cooling), above) are also provided in film form. Films use a bonding temperature of from 250°F to 325°F under a bonding pressure of 50 psi. Nonstructural bonding of metals, plastics, and cloth is possible where flexibility is desired.

Some nylons can be used alone as film adhesives. High melting point, high viscosity, and thermoplasticity are disadvantages, although some strength can be developed. The addition of epoxy corrects many of the drawbacks and produces what is in most ways the best structural adhesive yet developed. Their tensile strength of 7000 psi and peel strength of 90 lb/in are the highest available in structural adhesives. Low temperature performance is good, down to the cryogenic range. The disadvantages are: reduced strength above 180°F and susceptibility to moisture (both in the uncured film and the cured bond line). Cost is high for a film adhesive.

Uses are primarily within the aerospace industry in producing high strength, lightweight assemblies.

Polyolefins

Polyethylene and copolymers with polypropylene are used as protective coatings in bread wrap. Adhesion is low for these polymers. One practical method of improving surface adhesion of or to these films is by corona discharge treatment which forms an oxidized surface. An ethylene/acrylic acid copolymer also presents an oxidized surface. Both methods give greatly improved adhesion.

Polyvinyl Butyral

As a film type adhesive, this polymer has found a niche as the interlayer in laminated safety glass. A 30-mil thickness is used in automobile windshields.

The vinyl butyral polymer is a good base adhesive but requires additional flexibility and moisture resistance for bonding glass. This is provided by a phenolic or polyvinyl ether modifier plus a high percentage of a plasticizer. The bond to glass will have a lap shear strength of from 2500 to 3000 psi.

Polyvinyl Chloride (PVC)

This is the third largest family of plastics (after polyethylene and polystyrene). Adhesive uses include solutions for film lamination, plastisols, and organosols which are special dispersions, powders, and films. Thermoplastic powders may be used as a binder in nonwoven fabrics.

PVC films are used to bond PVC to fabric (note the frequent match of substrate and adhesive type). Dielectric heating is used to form "stitched" bonds in automobile door panel assemblies (53).

PRESSURE SENSITIVE TYPES

A pressure sensitive adhesive may be defined as a substance capable of holding things together when the surfaces are brought into contact under briefly applied pressure at room temperature. Generally, pressure sensitive adhesives are composed of a rubbery type polymer which is modified with several additives such as plasticizers, tackifiers, fillers, solvents, and stabilizers. In order of decreasing volume and, generally, increasing price, the following are used as basic polymeric raw materials: natural rubber, styrene-butadiene rubber, reclaimed rubber, butyl rubber, and butadiene-acrylonitrile rubber, polyacrylates, and polyvinyl alkyl ethers.

The rubbers as a whole are used in pressure sensitive adhesives because of good flexibility, ease of tackification and probably most important, economics. A major disadvantage of the rubbers is poor aging which results in a loss of elasticity and tack and a yellowing of the adhesive layer. The acrylates and polyvinyl alkyl ethers are more expensive but exhibit better aging characteristics. Thus, when rigid, high performance standards are required for a pressure sensitive adhesive, polyvinyl alkyl ethers or polyacrylates become the polymers of choice.

Millions of pounds of natural rubber are consumed for pressure sensitive adhesives. Most of this amount is used for masking tape and labels. In addition, large amounts of reclaimed rubber and styrene-butadiene rubber are used alone or in blends with natural rubber or polyisoprene for tapes and labels with a variety of backing materials. The acrylates and polyethers are used to a smaller extent for similar applications. Paper, cellophane, plastic films such as polyethylene, polyester, polyvinyl chloride, cloth and metal foils may be used as backing for such tapes. Generally, the adhesive coat is applied by calendering or solvent spreading. Tapes are made both with adhesive on one face and with adhesive on both sides (double-faced). The one face tape is used for sealing, masking, holding, and reinforcing operations. The double-faced tapes are used primarily for joining operations where there is a need to control the amount of adhesive and where pressure sensitivity is required.

Another major end use for pressure sensitive adhesives is in surgical tapes. The most important polymer for this application has been natural rubber which is often blended with tackifiers, oils, and pigments, but new developments using synthetic elastomers are changing the production picture.

NEW TECHNOLOGY

One-Component Systems

One of the objectives of the adhesive formulator is to be able to convert his two-package system into a stable one-package system, curable at room temperature. Moisture activated systems have already been described. An alternate successful route is the anaerobic adhesive based on acrylate diesters. Air inhibits the curing mechanism, so the system remains liquid until sealed off from air. The accelerator is activated by contact with metal. Thus the adhesive works exceptionally well when applied as a "liquid lock washer" for bolts and screws.

Microencapsulation of hardeners is related to adsorption by molecular sieves, but is not yet practical. What is needed is an inert, impenetrable but crushable shell surrounding a small droplet of the liquid hardener. Problems of leakage and short shelf life in the mixed system must be overcome.

Curing and Pretreatment Techniques

Radiation cure of vinyl type monomers adsorbed in wood is gaining impetus. The use of a radioactive source in place of heat to cure glue lines should find practical applications. Dielectric curing also is coming into use.

Treatment for hard-to-bond plastics include flame treatment, surface sulfonation, or corona discharge to provide a more active surface.

The "CASING" concept—crosslinking by activated species of inert gases—is reported to strengthen the inherently weak surface layer of plastic films (54). This gives a better surface for bonding without altering electrical and chemical properties, etc. which the corona discharge or other treatments do affect.

As the technology develops, adhesives will become stronger, cover a wider range of use, and will be simpler to apply. The manufacturer will become increasingly more capable of solving his production problems through adhesives.

REFERENCES

1. Henry Lee and Kris Neville, *Handbook of Epoxy Resins*, 2nd ed. (New York: McGraw-Hill Book Company, Inc., 1967).
2. G. J. Barenholtz, et al., "Correlation of Some Thermal Properties of Epoxy Adhesives and Shear Strength of Steel Bonds," *Industrial and Engineering Chemistry—Product Research and Development*, Vol. 2, No. 4 (December, 1963), pp. 270–2.
3. H. Dannenberg, "Determination of Stresses in Cured Epoxy Resins," *SPE Journal*, Vol. 21 (July, 1965), pp. 669–95.
4. R. R. Irving, "Adhesives Bonding; Its Day is Coming," *Iron Age*, Vol. 201 (April 11, 1968), pp. 83–90.
5. R. W. Drisko, et al., "Testing Adhesion of Epoxies that Cure Underwater," *Materials Production*, Vol. 4, No. 9 (September, 1965), pp. 24–8.
6. N. A. Debruyne and R. Houwink, *Adhesion and Adhesives*, Chapter 5C (New York: Elsevier Publishing Company, 1951).

7. J. M. Black and R. F. Blomquist, "Metal-Bonding Adhesives for High-Temperature Service," *Modern Plastics,* Vol. 33, No. 10 (June, 1956), pp. 225–8, 230, 235–6.
8. "New Adhesive System Saves Time, Money," *Chemistry and Engineering News,* Vol. 46, No. 18 (April 22, 1968), pp. 40–1.
9. J. R. Lawrence, *Polymer Resins* (New York: Reinhold Publishing Corporation, 1960).
10. P. J. Trent, et al., "Test Variables in Evaluating Chemical Resistance of Polymers," *Modern Plastics,* Vol. 43, No. 8 (April, 1966), pp. 172–4, 177, 179–80, 182, 222–4.
11. A. Damusis, *Sealants,* Chapters 6 and 7 (New York: Reinhold Publishing Corporation, 1967).
12. G. Schmutzler "Liquid Thioplastics and Their Application Possibilities," *Technik,* Vol. 19, No. 8 (1964), pp. 530–32.
13. B. A. Dombrow, *Polyurethanes,* Chapter 7, 2nd ed. (New York: Reinhold Publishing Corporation, 1965).
14. J. Delmonte, "Bonding Thermoplastics with Improved Urethanes," *Adhesives Age,* Vol. 11 (April, 1968), pp. 27–8.
15. S. L. Reegen, "Peel Strength of Polyurethanes from Oxypropylene Polyols—I: Polyol Structure and Fillers," *Journal of Applied Polymer Science,* Vol. 9, No. 1 (January, 1965), pp. 279–94.
16. ———, "Peel Strength of Polyurethanes from Oxypropylene Polyols—II: Effect of Additives, Isocyanate Structure, and Other Factors," *Journal of Applied Polymer Science,* Vol. 10, No. 9 (September, 1966), pp. 1247–59.
17. U. S. Patent No. 2 829 984 (April, 1968).
18. A. Damusis, *Sealants,* Chapter 4 (New York: Reinhold Publishing Corporation, 1967).
19. S. Sterman and J. B. Toogood, "How to Promote Adhesion with Silicones and Silanes," *Adhesives Age,* Vol. 8, No. 7 (July, 1965), pp. 34–39.
20. M. J. Bodnar, ed., "Structural Adhesive Bonding," *Applied Polymer Symposia No. 3* (New York: Interscience Publishing Company, 1966), pp. 369–95 and 455–88.
21. *Ibid.,* pp. 455–69.
22. H. A. Burgman, et al., "Aromatic Imide Polymers for High-Temperature Adhesives," *Journal of Applied Polymer Science,* Vol. 12 (April, 1968), pp. 805–29.
23. C. Clement Anderson, "Adhesives," *Industrial and Engineering Chemistry,* Vol. 59, No. 8 (August, 1967), pp. 91–96.
24. S. Livtak, "Polybenzimidazole Adhesives for Bonding Stainless Steel, Beryllium and Titanium Alloys. Part I: Adhesive Performance of PBI," *Adhesives Age,* Vol. 11 (January, 1968), pp. 17–24.
25. ———, "Polybenzimidazole Adhesives for Bonding Stainless Steel, Beryllium and Titanium Alloys. Part II: Adhesive Sandwich Study," *Adhesives Age,* Vol. 11 (February, 1968), pp. 24–28.
26. M. J. Bodnar, ed., "Structural Adhesives Bonding," *Applied Polymer Symposia No. 3* (New York: Interscience Publishing Company, 1966), pp. 279–318.
27. U. S. Patent No. 3 372 144 (March 5, 1968).
28. J. A. Clarke and J. M. Hawkins, "Characteristics of a Urethane Modified Structural Epoxy Adhesive," *Preprints of Division of Organic Coating and Plastics Chemistry,* Vol. 28, No. 1, American Chemical Society 155th Meeting (April, 1968), pp. 468–75.
29. D. J. Conway, "Metal-to-Metal Adhesives for Structural Applications at Elevated Temperatures," *Adhesives Age,* Vol. 11 (May, 1968), p. 24.
30. D. E. Floyd, *Polyamide Resins,* 2nd ed. (New York: Reinhold Publishing Corporation, 1966).
31. *Adhesive Bonding Alcoa Aluminum* (Pittsburgh: Aluminum Company of America, 1967), p. 90.
32. M. J. Bodnar, ed., "Structural Adhesives Bonding," *Applied Polymer Symposia No. 3* (New York: Interscience Publishing Company, 1966), pp. 1–25.
33. B. S. Gorton, "Interaction of Nylon Polymers with Epoxy Resins in Adhesive Blends," *Journal of Applied Polymer Science,* Vol. 8, No. 3 (May, 1964), pp. 1287–95.
34. W. M. Smith, *Vinyl Resins,* Chapter 5 (New York: Reinhold Publishing Corporation, 1958).
35. F. P. Baldwin, et al., "Elastomeric Prepolymers for Adhesives and Sealants Provide Improved Strength and Versatility," *Adhesives Age,* Vol. 10, No. 2 (February, 1967), pp. 22–29.
36. I. Skeist, *Encyclopedia of Polymer Science and Technology,* Vol. 1 (New York: John Wiley & Sons, Inc., 1964), pp. 483–92.
37. C. A. A. Rayner, *Adhesion and Adhesives,* Vol. 1, edited by R. Houwink and G. Salomon (Amsterdam: Elsevier Publishing Company, 1965), p. 329.
38. H. W. Coover, Jr., *Handbook of Adhesives,* edited by I. Skeist (New York: Reinhold Publishing Corporation, 1962), p. 409.
39. "Do Hot Melts Need an Apology?" *The Thermogrip Applicator,* No. 8, USM Corporation (October, 1966), pp. 2–3.
40. D. E. Truax, "Hot Melt Adhesives and Their Uses," *Tappi,* Vol. 49 (May, 1966), pp. 100A–2A.
41. C. C. Anderson, "Adhesives—Annual Review," *Industrial and Engineering Chemistry,* Vol. 59, No. 8 (August, 1967), p. 93.
42. J. W. Sawyer and R. E. Stuart, "Ionomer Laminates—New Plastics/Metal Composites," *Modern Plastics,* Vol. 44, No. 10 (June, 1967), pp. 125–8.
43. D. E. Floyd, *Polyamide Resins,* 2nd ed. (New York: Reinhold Publishing Corporation, 1966).
44. L. R. Vertnik, "Bonding with Polyamide Adhesives," *Adhesives Age,* Vol. 10, No. 8 (July, 1967), pp. 27–29.

45. P. Weitz, "Polyvinyl Methyl Ether—A Novel Route to Improved Adhesion." *Paper presented at Adhesive and Sealant Council Spring Seminar*, 1966.
46. R. N. Traxler, *Asphalt* (New York: Reinhold Publishing Corporation, 1961).
47. J. F. Clarke, et al., "Review of Ceramics—Metal Joining," *AFML-TR-65-143* (May, 1965).
48. R. C. Kausen, "Adhesives for High and Low Temperatures—I," *Materials in Design Engineering*, Vol. 60 (August, 1964), pp. 94–98.
49. E. F. Hess, "Adhesives for Fabricating High Strength Structures," *Metal Progress*, Vol. 83 (June, 1963), pp. 97–100, 120–8.
50. W. H. Smarook and S. Bonotto, "Adhesion of Carboxyl—Containing Polyolefins," *Journal of Polymer Engineering and Journal of Polymer Science*, Vol. 8 (January, 1968), pp. 41–49.
51. J. W. Sawyer and R. E. Stuart, "Ionomer Laminates—New Plastics/Metal Composites," *Modern Plastics*, Vol. 44 (June, 1967), p. 126.
52. J. G. Kaufman, "Fracture Toughness of 7075-T6 and T651 Sheet, Plate, and Multilayered Adhesive—Bonded Panels," *Journal of Basic Engineering* (September, 1967), pp. 503–7.
53. "Electronic Heat-Sealing Keeps Up to Date," *Modern Plastics*, Vol. 44 (July, 1967), pp. 86–9, 148.
54. R. H. Hansen and H. Schonhorn, "A New Technique for Preparing Low Surface Energy Polymers for Adhesive Bonding," *Journal of Polymer Science*—Polymer Letters Part B, Vol. 4, No. 3 (March, 1966), pp. 203–9.

chapter 2

design considerations

When joining materials with adhesives, initial design considerations must include not only the physical and chemical characteristics of the adherends and adhesive but also the operational environment that the bonded parts will encounter in use. Maximum bond integrity can only be realized when everything in the joining operation is carefully examined, particularly those factors related to the surfaces of the materials being bonded (1, 2). Major chemical and physical phenomena that must be considered in establishing a strong joint are:

1) Porosity and surface condition of the adherends
2) Surface tension of the adhesive and "wettability" of the materials being bonded
3) Physical properties of the adherends and the cured adhesive, i.e., tensile strength, shear strength, compressive strength, modulus of elasticity, thermal coefficient, etc.
4) Rheological and physical characteristics of the adhesive, i.e., viscosity, homogeneity, film thickness
5) Chemical compatibility of the adhesive and adherend, i.e., residual chemical groups, and pH of the adhesive
6) Methods of applying and curing the adhesive
7) Formation and removal of volatile by-products during cure.

The surface condition of the adherends themselves is most important in the formation of a strong adhesive bond. From a practical viewpoint all solid surfaces are rough and, when viewed under a microscope, appear to consist of hills and valleys and sometimes craters. These surfaces have moisture, gases, and other substances on them which are adsorbed from their surrounding environments. By displacing these adsorbed substances from the surface of the adherends, a properly formulated adhesive can usually form a strong tenacious bond. In addition to being rough, however, a surface can also be porous, and the degree of porosity varies greatly from one material to another. Examples of porous solids include wood, paper, leather, rubber, cork, fabrics, foams, and compressed metal powder parts. Because of the wide variations in porosity, surface conditions, and chemical characteristics of these surfaces, care must be exercised in the selection and application of the adhesive. Particular attention must be given to surface preparation, for failure of bonded joints usually occurs at the interface between adhesive and adherend (2).

In the classical adhesive bonded joint there are five separate components that must be considered: first adherend, first interface, adhesive film, second interface, and second adherend (3). If a suitable selection of adherends and adhesive has been made, if the surfaces of the adherends have been cleaned and prepared correctly, and if the adhesive has been applied by the recommended procedures, a high strength joint should result. Maximum performance can only be achieved, however, by properly engineering construction of the joint to withstand the various environments and stresses to which it will be subjected.

With these concepts in mind, it is readily apparent that the design and bonding of high performance adhesive joints is not a simple matter. In the following sections, the various elements of joint design are presented and discussed in detail.

ADVANTAGES AND DISADVANTAGES OF ADHESIVE BONDING

Each year millions of pounds of adhesives are used in the automotive, aircraft, building construction, marine, aerospace and other industries. The use of adhesives will undoubtedly continue to grow at a rapid rate. In designing joints, the many advantages of joining structures with adhesives must always be weighed against the limitations of the disadvantages.

Advantages

1) Distributes stresses uniformly over a large area eliminating localized stress concentrations
2) Joins similar and dissimilar materials
3) Provides capability to bond very thin and fragile materials
4) Joins materials of all sizes and shapes
5) Produces joints with smooth contours
6) Seals joints against a variety of environments, insulates against heat, and prevents electrolytic corrosion
7) Dampens vibrations, resists thermal shock and impact loads, and improves fatigue life
8) May be cured at relatively low temperatures eliminating the high temperatures required in welding and brazing
9) Frequently less expensive than other fastening methods
10) Can provide a significant weight savings
11) Simplifies joint design with a variety of adhesives available.

Disadvantages

1) Requires careful selection of adhesive to resist specific environments
2) Demands rigid attention to surface preparation and cleanliness
3) May require heat and pressure or a relatively long period of time to cure the adhesive
4) Bonding fixtures and jigs are generally necessary
5) Upper service temperatures are low compared to many metals
6) Non-destructive inspection of joints is difficult

7) Bonded joints cannot be readily disassembled
8) Rigid process control is generally needed.

The design engineer should be familiar with all operations involved in the joining process. He must understand the advantages and limitations of adhesive bonding, be familiar with the more conventional methods of joining, have a good grasp of economic factors and production practices, and be able to determine performance and material requirements as well as joint geometry. With these general concepts in mind, he should be aware that the bonding of high performance adhesive joints is not a simple matter but one of study and judicious understanding of the various requirements that help to make a good joint.

MATERIALS TO BE BONDED

Today, design and manufacturing engineers in all industries have a wide variety of materials available for the construction of bonded assemblies. Some of these materials are thin and fragile and can only be bonded with adhesives. Many high strength materials can be joined with adhesives more economically and out-perform assemblies made by older joining methods. Adhesive bonding is particularly advantageous for joining porous materials and nonmetallics and for bonding dissimilar substances. In many instances, adhesives provide the only satisfactory means of joining materials.

Several general operational parameters should be kept in mind when bonding materials with adhesives:

1) All parts must be clean and structurally sound
2) The adhesive employed should not seriously degrade either substrate.

Similar Substrates

Generally, it is easier to select an adhesive and design a bonded joint when the adherends are identical or similar. It is easier to bond two pieces of wood, for instance, than to bond wood and polyethylene. In all cases, however, the nature of the surface, porosity, cleanliness, and surface preparation must be carefully considered before attempting to join materials or assemblies with adhesives (see Chapter 4).

Porous Materials. Porous materials that are frequently bonded with adhesives to the same or similar substrates are wood, paper, rubber, fabrics, leather, foamed plastics, some ceramics, and porous metals. Porosity of a substrate or a surface may vary considerably with any type of material. Wood, for instance, is a relatively porous material, and adhesives for wood must be formulated to prevent excessive absorption into the adherend. When selecting adhesives for wood, consideration must be given to such variables as moisture content, percent of extractables, surface effects, density, hardness, and strength. The more dense hardwoods such as oak, maple, and walnut, form somewhat stronger adhesive joints than can be obtained with the less dense softwoods such as cedar, pine, and fir. As porosity increases, the possibility of forming starved glue lines increases because the adhesive tends to be absorbed into the pores.

Bonding of porous surfaces requires special handling to prevent absorption of the adhesive and formation of weak joints. Some preventive measures that are usually effective are:

1) Use high viscosity adhesives
2) Formulate with high molecular weight resins and polymers
3) Maintain high solids content
4) Bond or cure with minimum pressure application
5) Incorporate fillers into the adhesive
6) Partially react adhesive before application to substrate and before application of pressure.

Porous adherends are particularly helpful in the elimination of residual solvent or volatiles produced during cure. Paper is frequently bonded with adhesives based on polyvinyl acetate dispersed in water; the paper absorbs the water leaving the adhesive on top of the paper ready for bonding. Adhesives based on phenol-formaldehyde, resorcinol-formaldehyde, urea or melamine resins produce volatiles through chemical reaction during cure. These volatiles escape easily through most wood adherends but could cause blisters in metal to metal bonding if special precautions are not taken.

Nonporous Materials. From a practical standpoint the majority of metal substrates can be considered as nonporous materials except for a few materials such as compressed metal powders or metal foams.

Other examples of nonporous materials would include laminates of various types, i.e., epoxy-glass fabric, phenolic-canvas, melamine-paper, etc., cast, extruded or molded plastic sheets, molded polyester-fiberglass, and glass. The usual procedure in bonding materials of this sort is to select an adhesive as nearly like the substrate as possible. Modified epoxy adhesives are frequently employed on reinforced thermosetting resinous materials. Most thermoplastic substrates require special adhesives and/or special surface treatments in order to produce satisfactory bonds. Since the elimination of volatiles is frequently a problem in bonding nonporous materials, many fabricators include glass fabric, chopped glass or other fillers in the adhesive to provide an escape path for the volatiles, to enhance bond strength, and to control bond line thickness.

Dissimilar Substrates

Special consideration must be given to the design of bonded joints of dissimilar materials. When different metals are involved, the surface preparation for each metal must be carefully followed. Consideration must be given to the different coefficients of thermal expansion for all the involved materials. Many times a single adhesive system will not bond two dissimilar substrates, and various primers may have to be applied to each adherend before bonding with adhesives. A duplex adhesive film can also be used to bond two dissimilar adherends.

Porous Materials. When attempting to join dissimilar porous surfaces such as rubber and porous metal or rubber and wood, special bonding systems are frequently required. Modified rubber adhesives can also be used on both substrates, depending upon the application. Wood may have to be filled, sealed, or primed before application of the adhesive, and the rubber may require cyclizing. The general procedure for bonding porous materials outlined at the start of the chapter must be followed. Wood can be bonded to metal, for instance, with a neoprene adhesive. Polystyrene foam can be bonded to metals with a modified epoxy, to metal foils with polyvinyl acetate, or to paper or plywood with vinyl or rubber base adhesives. In all cases, the formulation and application of the adhesive must be such that it does not flow off the surface and/or into the pores.

Nonporous Materials. Examples of the bonding of dissimilar nonporous materials can be readily illustrated in the bonding of metal to rubber and metal foils to laminates or plastics. A strong bond can be formed between rubber and metals using isocyanate and rubber base adhesives. In some cases with vulcanized rubbers, it has been necessary to cyclize the rubber with acids or prepare the rubber surface by chlorination. Metal foils such as copper are readily bonded to phenolic-paper and epoxy-glass fabric laminates through the use of phenolic-polyvinyl acetal adhesives. Even some rubber base adhesives have performed well in this application. Thin metal foils, of course, have been bonded for many years to plastics using hot melt and polyvinyl acetate type adhesives.

AREA SIZE TO BE BONDED

Joint design aims at obtaining maximum strength for a respective area. The selection of a joint design can be determined by limitations in production facilities, production costs, or the desired final appearance of the part. The strength of an adhesive joint is determined primarily by the mechanical properties of the materials involved, residual internal stresses, degree of true interfacial contact, and the geometry of the joint. Failure to pay close attention to any one of these factors can result in poor joint performance. Elimination of stress concentrations is of major concern to the design engineer since such stresses reduce the strength and useful life of the joint. Unfortunately, these localized stresses are not always apparent and may occur as a result of differential thermal expansion of the adhesive and adherends or from the shrinkage of the adhesive during the cure cycle. Volatiles trapped in the adhesive may also set up internal stresses, particularly at the interface. Internal stresses decrease as the thickness of the adhesive layer decreases, and the tendency to trap volatiles in the adhesive layer is reduced. Entrapment of air at the interface can occur if the adhesive has too high a viscosity, does not flow easily as it is cured, or if it does not wet the substrate.

An adhesively bonded joint can be loaded in compression, tension, shear, cleavage, or peel. When loaded in pure compression, a joint is less likely to fail than when loaded in any other fashion, but such joints are limited in application. Structural adhesives are usually strong in shear and weak in cleavage and peel, and adhesive bonded joints are designed with these characteristics in mind. Loading a joint in pure tension is generally impractical because, under high stress, joint components can easily get out of line or the applied forces can become misaligned resulting in generation of the relatively weak cleavage and peel forces. The usual practice in designing adhesive joints is to minimize the peel and cleavage forces by placing the adhesive in shear and/or compression for maximum strength and applying a thin, uniform adhesive layer over the largest practical bond area.

JOINTS

Corner Joints—Sheet Metal

Corner joints can be assembled with adhesives by using simple supplementary attachments. This permits joining and sealing in a single operation.

Typical designs are slip joint (see Fig. 2–1), right angle butt joint (see Fig. 2–2), and right angle support joint (see Fig. 2–3). These joints give the structure an increased degree of rigidity. Adhesives with void filling properties, such as modified epoxy types, are most frequently used. The possibility of using a heat curing epoxy depends on the heat resistance of the materials being bonded.

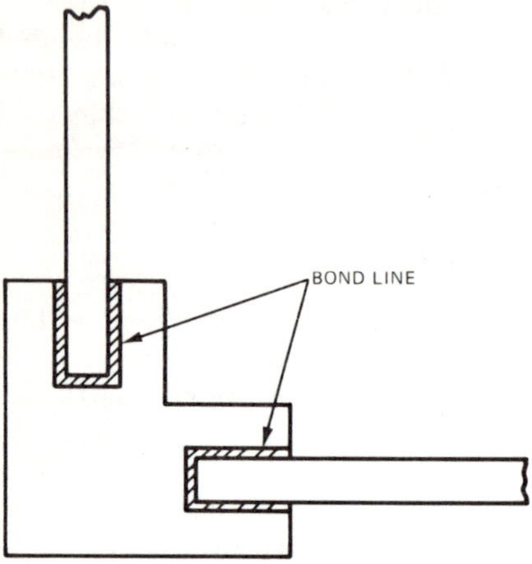

Fig. 2–1. Slip joint.

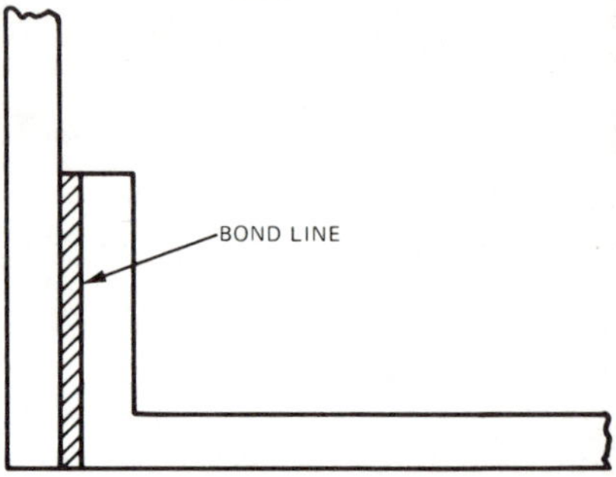

Fig. 2–2. Right angle butt.

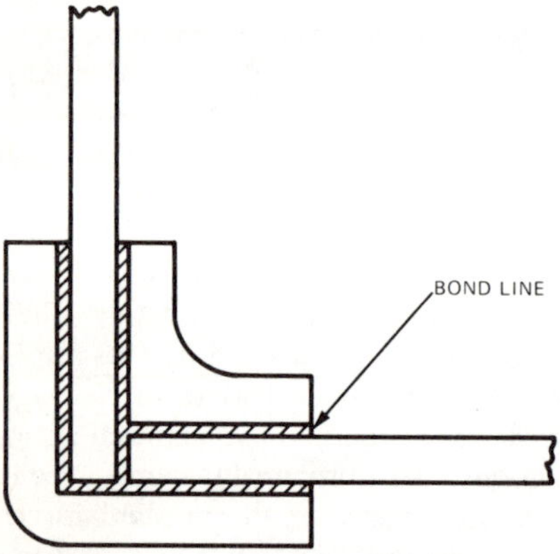

Fig. 2–3. Right angle support.

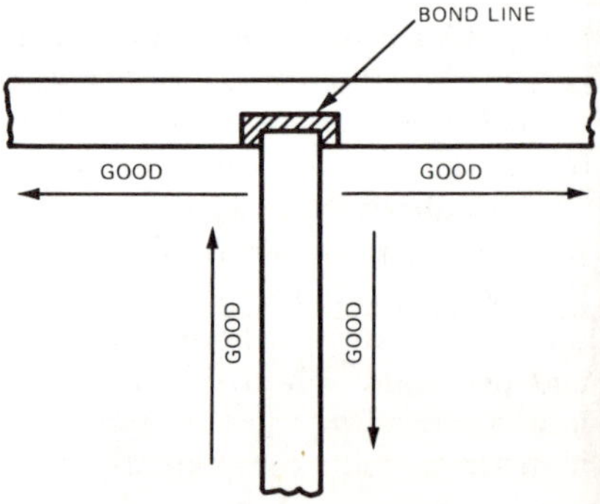

Fig. 2–4. Angle joint.

Thin gauge metals or sandwich panels can be easily joined with a heat curing epoxy to form boxes or cases, housings, vehicle bodies, metal boat hulls, etc. (4).

Angle Joints

Angular joints (see Figs. 2–4 through 2–6) give rise to either peel or cleavage stresses depending on the gauge of metal used. Typical approaches to the reduction of cleavage are illustrated (4).

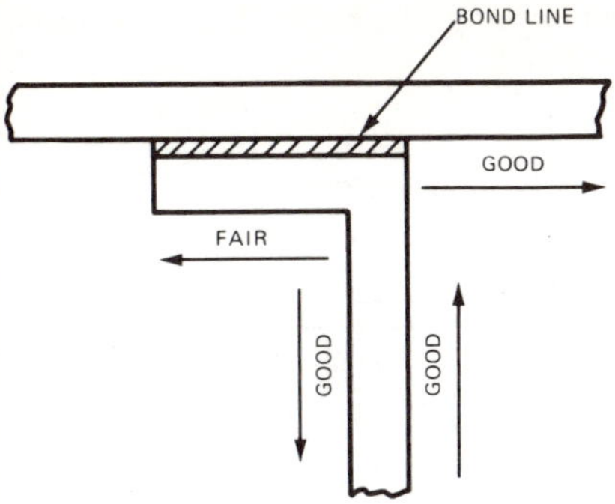

Fig. 2–5. Angle joint.

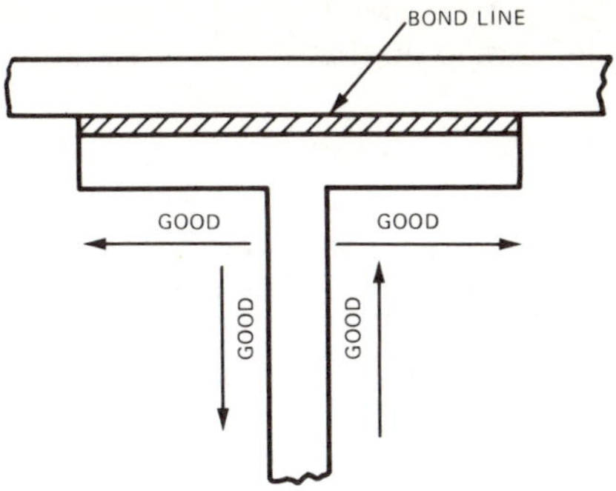

Fig. 2–6. Angle joint.

Butt Joints

A straight butt joint is weak in cleavage. Recessed butt joints, such as landed scarf tongue and groove (see Fig. 2–7), conventional tongue and groove (see Fig. 2–8), and scarf tongue and groove (see Figs. 2–9 and 2–10) are recommended. Landed scarf tongue and groove joints have an advantage of acting as stops which could be utilized to control adhesive line thickness. Tongue and groove are self-aligning during assembly and act as a reservoir for mastic type void filling adhesives (4).

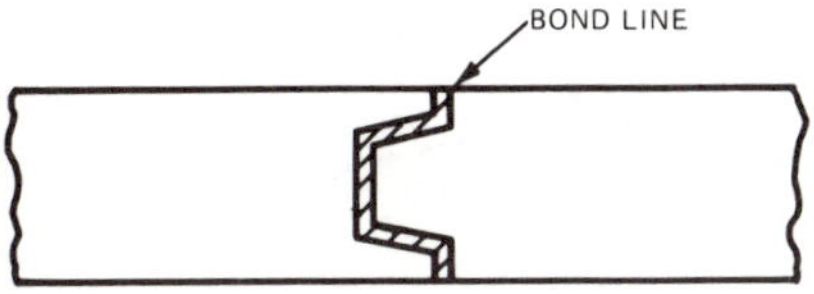

Fig. 2–7. Landed scarf tongue and groove.

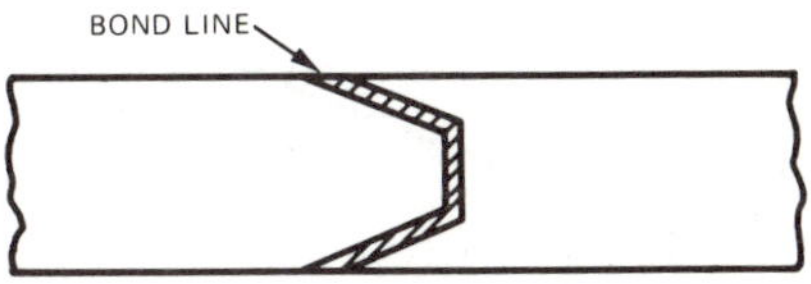

Fig. 2–8. Scarf tongue and groove.

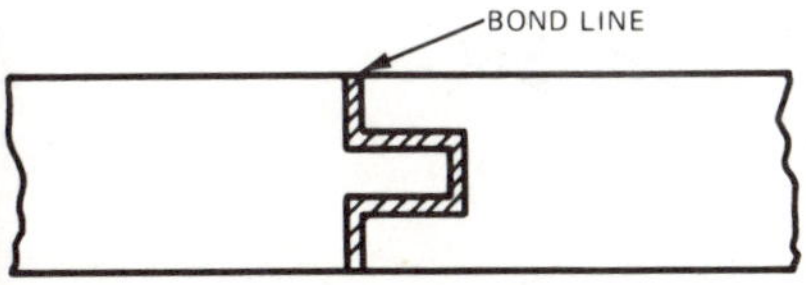

Fig. 2–9. Conventional tongue and groove.

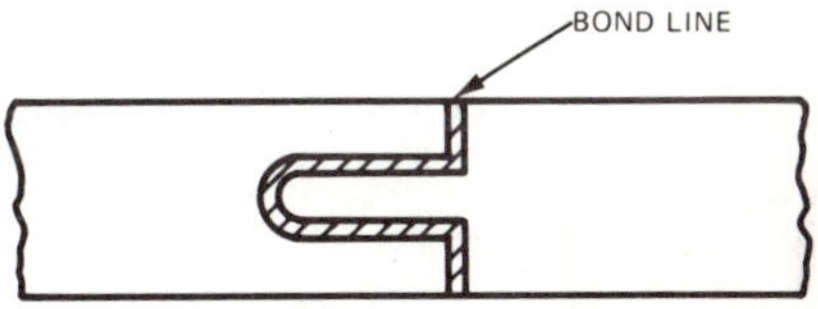

Fig. 2–10. Typical adhesive bonded butt joint.

Stiffener Joints

Deflection and flutter of thin metal sheets can be minimized by the use of adhesive bonded stiffeners. When such assemblies are flexed, peel stresses are exerted on the adhesives. If the flanges on the stiffening section can bend with the sheet, minimum peel stress on the bond will result. Increasing the gauge of the sheet or decreasing the gauge of the stiffener flange will give equivalent results. Several types of stiffening members, such as T-sections (see Fig. 2–11), hat sections (see Fig. 2–12), and corrugated backing (see Fig. 2–13) can be used depending on the degree of rigidity desired (4).

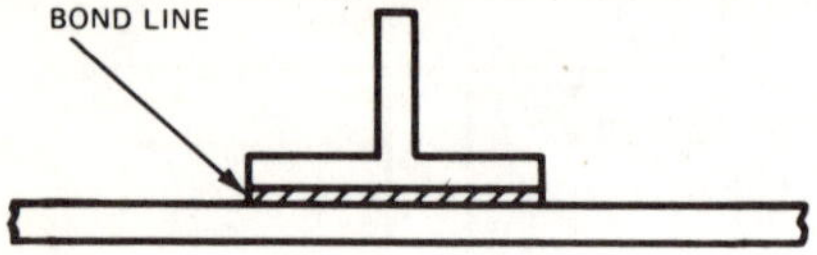

Fig. 2–11. T section.

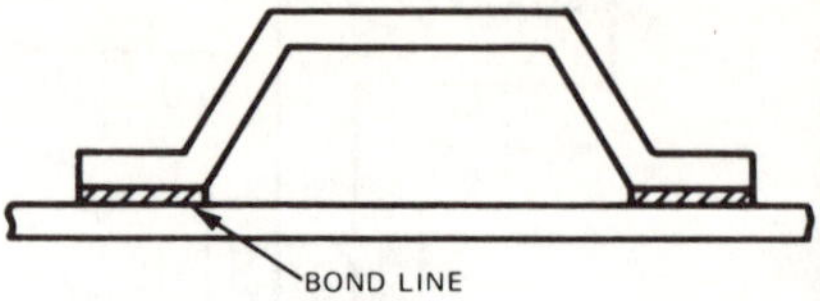

Fig. 2–12. Hat section.

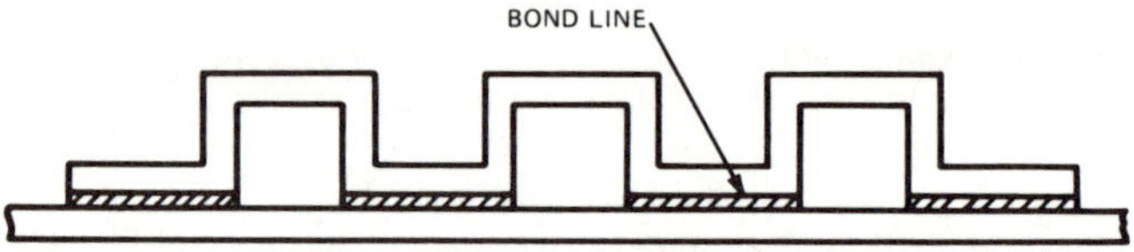

Fig. 2–13. Corrugated backing.

Corner Joints—Rigid Members

Corner joints, such as those encountered in storm doors or decorative frames, can be bonded with adhesives. Woodworking joint designs such as end lap (see Fig. 2–14), mortise and tenon (see Fig. 2–15), and mitered joint with

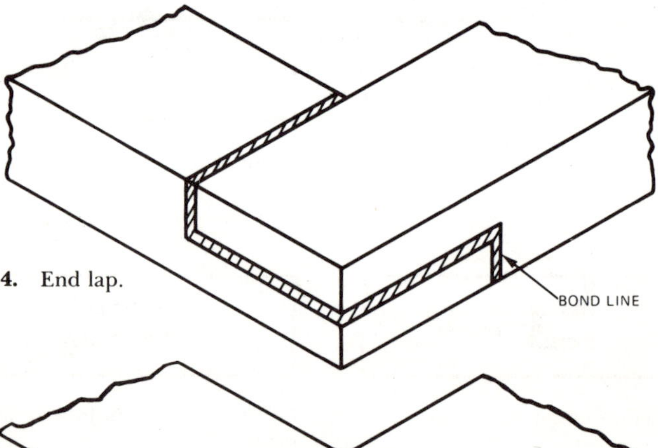

Fig. 2–14. End lap.

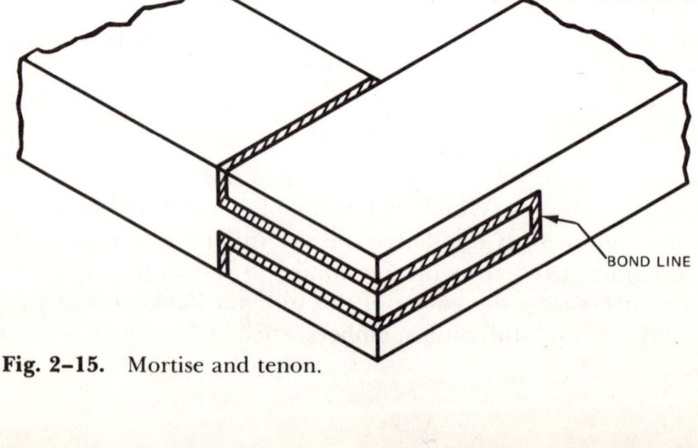

Fig. 2–15. Mortise and tenon.

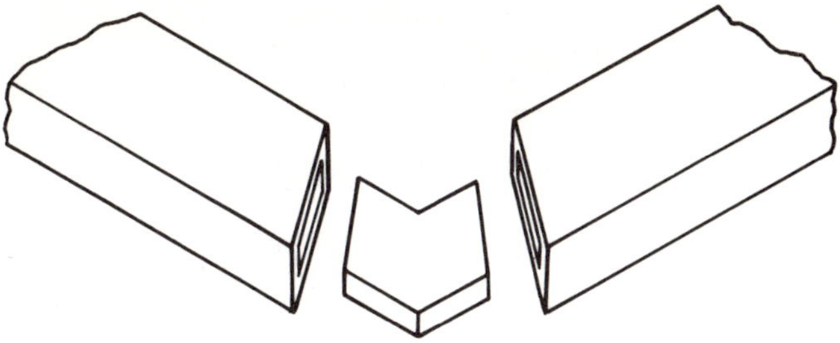

Fig. 2–16. Mitered joint with spline.

spline (see Fig. 2–16) are used to provide sufficient bonding area to resist inherent racking or twisting stresses. End lap joints are the simplest design type although they require machining. Adhesives requiring pressure during cure may be utilized in such designs. Mortise and tenon joints are excellent from a design standpoint but also require machining. A mitered joint with spline is best if members are hollow extrusions. In this case, a void filling modified epoxy adhesive is recommended (4).

Lap Joints

In adhesive joint design, the shear or lap joint is most generally used. Also, lap joints are the most practical and applicable to adhesive bonding of thin materials.

The simple lap joint (see Fig. 2–17) is off-set, and shear forces are not in line. This can result in the introduction of cleavage and peel stresses under load.

The joggle lap joint (see Fig. 2–18) is one way of providing stress alignment. It places the entire load-bearing area in the same plane as the shear stress on the adherends. In this type of joint, application of pressure for curing is easily accomplished, and the joint can be formed by simple metal forming operations.

A tapered single lap joint (see Fig. 2–19) is more efficient than a simple lap joint. The tapered edge allows bending of the joint edge when distortion occurs under stress.

The double butt lap joint (see Fig. 2–20) places the load-bearing area in the same plane as the shear stress on the adherends. This type of joint, however, requires machining which is not always feasible with thinner gauge metals.

Double scarf lap joints (see Fig. 2–21) have better resistance to bending forces

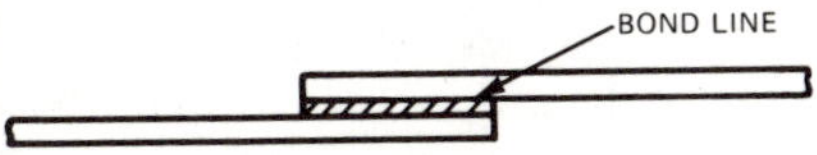

Fig. 2–17. Simple lap.

Fig. 2–18. Joggle lap.

Fig. 2–19. Tapered single lap.

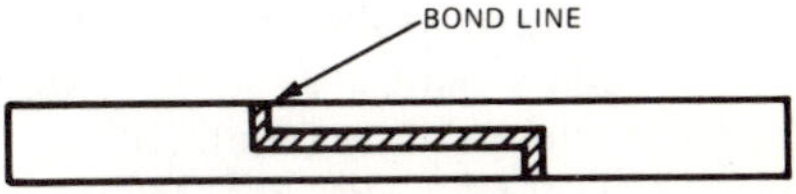

Fig. 2–20. Double butt lap.

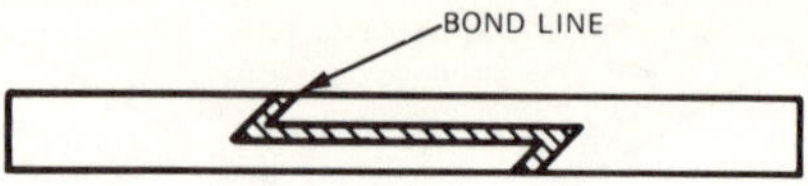

Fig. 2–21. Double scarf lap.

than double butt joints, but this type of joint also requires machining, increasing costs (4).

Honeycomb Sandwich Construction

In the aircraft industry honeycomb sandwich construction (see Fig. 2–22) is used extensively in the manufacture of control surfaces, bulkheads, flooring, doors, engine cowlings, helicopter blades, and wing panels. Present commercial applications include building wall panels, doors, house trailer flooring, small boat hulls, scaffolding, cargo containers, pallets, furniture and so on. The primary reason for using sandwich designs in preference to other construction are:

1) High strength-to-weight ratio
2) Fatigue resistance
3) High rigidity per unit weight
4) Surface smoothness
5) Insulation

Sandwich structures are produced by adhesively bonding rigid facings on both sides of a honeycomb core. Functionally, the sandwich structure can be compared to an I-beam where the facings resist tensile and compressive loads as do I-beam flanges; the honeycomb core simulates the web of the I-beam by carrying the shear stresses. In addition, however, the core provides continuous support for the facings, allowing their full strength to be utilized without buckling or crimping. It is the function of the adhesive in sandwich structures to transmit shear loads uniformly from the facings to the honeycomb core; good adhesive filleting is necessary in honeycomb bonding. For the completed panel to have optimum strength the adhesive must be distributed evenly between the core and the facings to form a T-joint. The most popular core-to-skin adhesives are based on epoxy resins because of their high strength and good wetting properties with only minimal volatiles produced during the bonding operation. Vinyl-phenolic and elastomer-phenolic adhesives are also widely used but are less likely to form dense fillets. Primers are frequently used with these latter systems.

Honeycomb cores can be made from many materials including steel, titanium and aluminum foils, glass reinforced plastics, films, and papers. The variety of available facing materials gives the engineer a considerable amount of flexibility in the design of sandwich structures. Selection of the proper core-to-skin adhesive is dependent upon the materials involved, the application, environmental exposures, service life, and weight considerations. Adhesives can be applied to core and facings with a spray gun or roller, but in structural applications adhesive films of controlled thickness are generally used. For many of these applications lightweight adhesives in the range of .025 to .080 lb/sq ft are used with curing cycles of 15 to 60 min at 225° to 250°F. Bonding pressures are usually in the 25 to 50 psi range, but depending upon the specific application, these pressures can vary from contact to 150 psi. While the emphasis has been on the lower curing temperatures, many modified epoxy and phenolic adhesives are formulated to cure around 350°F. Post-curing temperatures of 500° to 550°F are required in the case of polyimide adhesives.

Although sandwich structures are highly efficient, the manner in which loads are transmitted into and out of the sandwich largely governs service performance. Good design practice closes off the edges in sandwich panel construction to provide edge reinforcement for the transfer and distribution of edge attachment loads. Sandwich close-out and fastener design objectives are ease of fabrication, proper transmittal of shear loads and resistance to edge peel. As with any adhesive joint, stress concentrations are to be avoided and sandwich fasteners are so designed as to distribute loads evenly over a broad area. Prefabricated inserts and edging of wood, metal and reinforced plastic, epoxy potting compounds, and other materials can be used to provide the necessary strength and load distribution.

PERFORMANCE REQUIREMENTS OF THE JOINT

There are four primary considerations that must be taken into account when designing a joint with adhesives. They are: (1) strength, (2) service environment, (3) service life, and (4) weight. The effectiveness of the joint is usually affected by all four (5).

Strength

The first design consideration normally is strength. The amount of the load to be transferred needs to be known. The type and magnitude of the load to be transferred determines the size and shape of joint most feasible to carry the load. Designers should not ask the adhesive to carry "peel" loads unless the "peel" conditions are specifically known and the adhesive is tested for that specific condition. If "peel" loads might be introduced into the joint, "peel stopper" fasteners should then be used. The most efficient joints transfer loads in shear. Some typical joints used are shown in Figs. 2–1 through 2–21. Examples of sandwich-structured joints may be seen in Fig. 2–22.

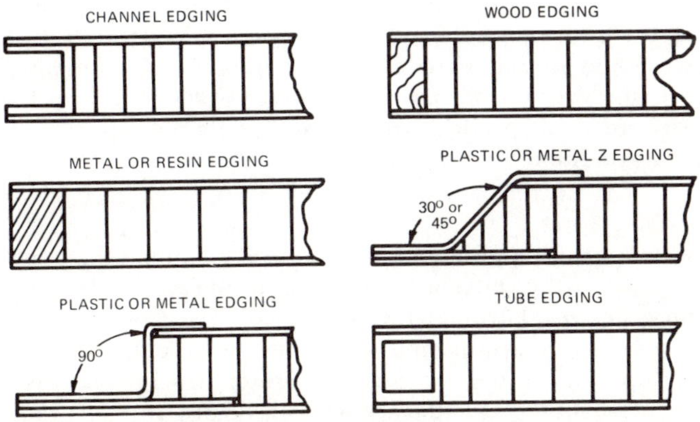

Fig. 2–22. Sandwich-structured joints.

Through comprehensive design analysis, most joints can be stress analyzed to determine their capabilities. Proof testing is sometimes more feasible and accurate. Often, both are needed to economically achieve the desired end product.

The allowable strengths used by design should be first established and verified as attainable through production processes and conditions. Often a designer will use reduced laboratory test value strengths only to find out, too late, that laboratory controls are not attainable under production conditions. The person who is responsible for "strength allowables" should recognize the production limitations when issuing information on allowable strengths for use in design. However, production people must recognize their responsibility to strive toward better controls in the bonding environment and systems.

Service Environment

When design criteria are being established, the responsible person should make sure that the information on allowable strengths also take into account the effect of the bonded joint's service environment on strength. Adhesive joints are commonly exposed to a service environment that includes: (1) temperature, (2) humidity, (3) temperature and humidity cycling, (4) solvents, (5) fuels, (6) oils, (7) light mechanical vibration, (8) impact, (9) noise vibration (acoustic fatigue), (10) oxidation, (11) corrosion, and (12) light stability. For example, brittle adhesives, such as phenolics, while good at elevated temperatures, are not normally good when impacted or highly vibrated at below 0°F temperatures. On the other hand, the more flexible polyamides are good at low temperatures and are poor at elevated temperatures above 180°F.

Relative to humidity, solvents, fuels, and oils, "sealing" of the bond joint does not necessarily eliminate the problem of environmental permeation especially in the case of perforated honeycomb. If adequate drainage around joints is not provided to preclude a "head" of liquid against a bond joint, most liquids will in time permeate the joint.

Service Life

Along with service environment, the service life of the joint being produced should be known. Under certain environmental conditions, such as elevated temperature, some adhesives deteriorate with respect to time. If exposure of a bonded joint to a particular environment for a known time is anticipated, the life expectancy of the joint can usually be determined through accelerated "aging" tests. Some aircraft parts have to be in service for an anticipated ten years or more whereas a missile part may be in stressed service for a few minutes. Most adhesive material specifications require qualification tests under various environmental conditions and exposure times.

Weight

In most aerospace applications, the desire to limit weight often necessitates changes from the conventional joining techniques and design to an adhesive bonded design. Through the use of adhesive bonded joints, thinner substrates can most often result in a lighter weight structure which has become increasingly important in the last decade. Lighter structure means: (1) higher performance capabilities, and (2) increased payloads. In end products where overall weight is not important, adhesive weight can still affect cost of adhesive (amount used), bond line thickness (strength of joint), and the need for accuracy in dispensing equipment.

Weight control is usually most easily attained by use of film adhesives unless exacting dispensing equipment is made available. Strength of the adhesive can affect the amount of adhesive needed to achieve the bonded joint. An example is honeycomb sandwich structure requiring high face-to-core tensile properties. The strength of an adhesive is the desired property that dictates how thick (or how much weight per unit area) it must be to achieve the joint strength needed. Given two adhesives at the same cost per pound but one is stronger in the desired property, the stronger adhesive in a lighter version can be less expensive provided the cost to achieve good fit of the joint does not become too high. Bond line thickness, which sometimes is best controlled by the amount of adhesive placed into a joint, also affects the bond strength.

Most companies involved with designing adhesive bonded structure develop their own methods of design analyses. However, adhesive and core suppliers more often can provide guides to extremely helpful methods of design analyses.

REFERENCES

1. J. J. Bikerman, *The Science of Adhesive Joints* (New York: Academic Press, 1961).
2. Louis H. Sharpe, "Assembling with Adhesives," *Machine Design*, August 18, 1966.
3. D. R. Smith, "How to Prepare the Surface of Metals and Non-Metals for Adhesive Bonding," *Adhesives Age*, Vol. 10, No. 3 (March, 1967), pp. 25–31.
4. "Scotch-Weld Brand Structural Adhesives," *3M Company*, St. Paul, Minnesota, 1967.
5. Charles V. Cagle, *Adhesive Bonding: Techniques and Applications* (New York: McGraw-Hill Book Company, Inc., 1968).

chapter 3

selection of adhesives

The selection of an adhesive for a given application should be done with the advice of a qualified adhesive engineer. Most adhesive suppliers furnish engineering service as well as the product which is ultimately selected. However, the choice of the proper adhesive from the six classes listed in Chapter 1 can be simplified if qualifying "ground rules" are established first.

Four basic requirements should be considered when analyzing an adhesive application.

1) The type of material to be bonded is important—whether the adherends are metal, wood, paper, fabric, leather, plastic, elastomer, or ceramic. Next, physical properties of the adherends such as size, flexibility, porosity, and heat distortion temperature should be considered. Finally, the selection of the material should be engineered around the weakest member of the system. Since the quality of the final assembly may be structural, the strength of the adherend itself is important.
2) All possible process or bonding procedures should be thoroughly investigated. The most simple process that will satisfy the final requirements should be selected. All assembly steps, such as surface preparation, adhesive application, and curing or setting methods should be reviewed with the performance requirements as a guide. Keep the procedure as simple as possible. Special equipment such as fixtures, ovens (drying and bonding), and applicators should be included in the evaluation.
3) Before an adhesive can be selected intelligently *all* service conditions must be considered. A stress analysis is necessary to determine if the bond is under tensile, shear, peel, or cleavage loading. The stress cycle is also important. Whether the type of loading is continuous or intermittent should also be considered. Consideration of environmental conditions such as heat, cold, water, humidity, chemicals, oil, solvents in many special applications is fundamental in the selection of adhesives. Actual service tests should be duplicated if possible. Aging should be a part of this evaluation.
4) As a final requirement, evaluate the economics of the system. The cost of the adhesive itself per unit is usually minor compared with the cost of the

process. The cost of labor should prohibit the purchase of cheap but hard to apply adhesives. Automated equipment will reduce the cost of assembly.

The problem of selecting an adhesive for a given application can now be studied using the above suggested screening methods.

PHYSICAL REQUIREMENTS

Choosing the proper adhesive from one of the six types in Chapter 1 will be discussed in the order of bond strength, curing or setting characteristics, adhesive workable life, clean up, controlled application conditions, and safety.

Bond Strength

Bond strength is the measure of performance. The specification should cover the requirements under environmental conditions. Any evaluation program should include the temperature range, moisture conditions, and any other conditions related to the performance of the assembly.

Chemical Reactive Types. The chemical reactive type adhesives are usually chosen for structural bonding where high strength over a wide range of temperature is required. In both the heat reactive or catalyst-cured adhesives, strengths from 2000 to 6000 psi are possible (see Table III–1). The choice of the resin or elastomer based material is dependent on the elasticity required in service. These materials may require heat and pressure fixtures during the curing cycle. Surface preparation for structural bonding is important. Many times the failure of an adhesive is due to poor surface preparation. The surface should be clean, and the strength of any primer or surface conversion treatment should be stronger than the adhesive.

Table III–1. Examples of Structural Adhesives – Aluminum Bonding.

Adhesive Type	*Temperature Cure*	*Pressure (psi)*	*Tensile (psi)*					*Peel (lb/in)*
			−67°F	*77°F*	*180°F*	*300°F*	*400°F*	
Modified Phenolic	350°F	100	1800	2200	2200	2400	2000	4
Epoxy Modified	350°F	Contact	3500	4000	3000	2500	2000	5
Epoxy	77°F	Contact	1300	3000	600	100		5
Epoxy Phenolic	350°F	100	2800	3000	3000	2400	1700	4
Vinyl Phenolic	350°F	100	3000	4200	2000	400		15
Nitrile Phenolic	350°F	100	3000	3800	2600	2300	1000	75
Neoprene Phenolic	200°F Activate	Contact	500	1000	100			25

Evaporation or Diffusion Types. Evaporation or diffusion adhesives achieve their strength by drying or crystallizing. The loss of the carrier, either solvent or water, leaves a dry film. The strength of the film is controlled by the ratios of elastomer, resin, plasticizer, and pigment. Evaporation type adhesive can be compounded to be used with any substrate. Since these compounds are not usually chemically reactive or cured, high constant loading may cause cold flow; their use is therefore limited to low load assemblies. Higher resin compound give less cold flow but tend to be brittle. Evaporation adhesives also show a higher degree of variation over a wide temperature range. Usually the maximum strength for these types of adhesives will be about 800 to 1000 psi at 80°F.

Hot Melt Types (Cooling). The strength of the hot melt adhesive is determined by the strength of the compounded polymers and resins. These are normally thermoplastic-based compounds with 100 percent solids and good cohesive strength. Application requires more specialized equipment than the evaporation type adhesive, but hot melts offer the advantages of having no solvent and a very fast set.

Delayed-Tack Adhesives. After a delayed-tack adhesive is heat activated and cooled, it will remain tacky anywhere from several minutes to several days over a wide temperature range even as the temperature cools to 50°F, but below 50°F, the adhesive will set.

Film Adhesives. Film type adhesives can be either thermoplastic or thermosetting. In either case, heat and pressure are normally used for the application. A few of the thermoplastic films may be used with solvent activation. The thermosetting films are used for high strength bonds over a wide temperature range. The bond strength from the thermoplastic film again is related to the tensile strength of the polymer. The softening range and melting point will dictate its applicable area. From an engineering standpoint, these are the simplest adhesives to apply.

Pressure Sensitive Type. As a rule pressure sensitive adhesives are not used in a permanent bond, especially where higher loading strengths are required. Their relatively low strength and high tack make them valuable where the substrate can be bonded, easily removed and rebonded. Pressure sensitive adhesives are ordinarily applied to one of the substrate materials from a solution or a hot melt. When substrates with large dimensional variations are bonded, a sealing tape extruded on a carrier may simplify the operation.

Setting Characteristics

The curing of a structural adhesive compound relies on polymerization or chemical reactivity with additive components. The reaction is accomplished by using a heat-time cycle or the utilization of a multiple component system. For heat activated adhesives, either solution or film type, the polymer or resin plus the curing agent are mixed in a single compound. Curing operation then requires an oven, platen press, autoclave, or some other heat source to complete the bond. A time-temperature curve to attain the optimum properties must be established for each application. This type of adhesive develops the highest bond strength.

Adhesives that cure at room temperature make use of a multi-component system. The polymer or resin is contained in one component, and the curing agent and catalyst are in the other. Proper mixing of the components will result in a compound that will cure at room temperature. Curing rates can be altered by varying the mix ratio or changing the catalyst. They are usually used where high strength is required, but the substrate does not lend itself to heat cycles. Generally, the room temperature cured adhesives do not have as wide a temperature latitude as the heat cured adhesives.

As pointed out previously, the evaporation type adhesives attain their strength from loss of solvent. The rate of evaporation controls the set time of the adhesive. Substrates are coated with the adhesive, and after the solvent evaporates, the coated surfaces are assembled. Hot melts reach their setting point on cooling; therefore, they must be used at temperatures below their softening point. Hot melt adhesives give a very quick bond. They work well on an automated line for fast assembly.

Adhesive Life

The application life of an adhesive must be considered from two angles; first, the shelf life or the aging characteristics of the compound and secondly, the pot life or the workable life of the adhesive.

Curing systems may advance in storage resulting in gelling plus loss of flow. Refrigerated storage will delay the reaction and increase the shelf life. This is found more often in the single component curing type. Usually the multiple component curing systems are stable; they will not react until they are mixed. Normally the shelf life of evaporation and hot melt adhesives is good. However, there are a few polymeric materials that will change even in solution; this is often reflected in viscosity growth. The manufacturer should be consulted for storage recommendations.

Pot life is an important factor in multiple component curing systems and in some evaporation adhesives. The pot life is dependent on the reactivity of the curing agent and catalyst. The workable life should be adjusted to production cycles.

The pot life of evaporation adhesives is not as critical as the curing type. Minor adjustments are made by varying the solvent. Slow drying solvents will increase the open time (see Glossary).

Cleanability

When selecting an adhesive, consideration of post application cleanup of the substrates and application equipment is important. Removal of excess adhesive on the bonded section could be difficult especially when curing compounds are used. Well designed application equipment will eliminate cleanup costs in most jobs.

Application Conditions

Ideal conditions would be to install a strict temperature and humidity control; for example, 77°F and 50 percent relative humidity. A controlled environment will eliminate many of the variables in adhesive application in areas such as spraying and roller coating. The variation in temperature has a decided effect on curing adhesives and the rate of evaporation of solution adhesives. The problem which often occurs in high humidity is "blushing" on the surface. "Blushing" is the condensation of atmospheric moisture at the bond line interface. When moisture is taken into the bond line, erratic bond strength results.

Safety Considerations

Whenever organic materials are formulated and used on the production line, a few precautions are always necessary. Some of the curing agents, catalysts, resins, and pigments are irritants and may cause discomfort. A few of the solvents are toxic and many are flammable or "red label." However, proper application equipment and ventilation reduces these hazards to a minimum.

When choosing an adhesive for an application, be sure to read the handling instructions. Very often, a careful examination of the manufacturer's label will point to the best method for its application. In cases where industrial adhesives are not labeled, the Purchasing or Engineering Departments usually have the specifications and literature on the adhesive for a choice of a method of application.

METHOD OF APPLICATION

Before choosing a method of applying the adhesive, the user should first consider the following:

1) Various methods for the best application of the adhesive
2) Economy of the application
3) Production rate for the piece being fabricated
4) Method of application from the point of view of simplicity
5) Best all around results.

As pointed out at the beginning of this chapter, the type of material to be bonded is important. The testing of the adherend should be initiated at the beginning of the investigation, but this testing should also be based primarily on the customers needs and wishes. The manner in which the adhesive responds to testing will often actually dictate the method of application. In most cases, however, the tests will reveal several suitable ways to apply the adhesive.

Methods of Adhesive Application Available (See also Chapter 5)

Adhesives may be applied manually by roller, brush, extrusion and flow, and trowel. They can also be applied semiautomatically with spray guns or high pressure extrusion guns, or they can be applied automatically by machine methods that are usually geared for mass production of a particular part. These methods are mentioned briefly at this point, for more information see Chapter 5.

Roller. Rollers for manual application can be constructed from wood, paper, cork, metal mesh, and synthetic fibers. The length of the nap normally varies with the amount of material needed as well as the final coated surface appearance.

Brush. Brush bristles are made of various types of hair, synthetic fibers, wood, and metal. A brush is used to manually apply an adhesive to a given surface. Some brushes are fed the adhesive by pressure.

Extrusion and Flow. Extrusion and flow are general methods of applications used by workmen operating caulking guns. Extrusion application is also used in curtain coaters, flow-coaters, and high pressure air powered extrusion units.

Trowel. A trowel normally leaves a fairly constant amount of adhesive on the work surface. This is accomplished by small cutouts along one margin of the trowel which allows a predictable amount of material to pass through and remain on the surface.

Airless Spray. The pressurizing of the adhesive through a small opening atomizes it. The spray pattern coats the surface to be sprayed (see Fig. 3–1).

Air Spray. Low fluid pressure moves the material to the spray tip area where air is introduced to atomize the adhesive into an acceptable spray pattern (see Fig. 3–1). The force to move the adhesive is usually derived from an air power pump of a pressurized diaphragm type pump.

High Pressure Extrusion. Fig. 3–2 illustrates a typical high pressure extrusion pattern on a surface.

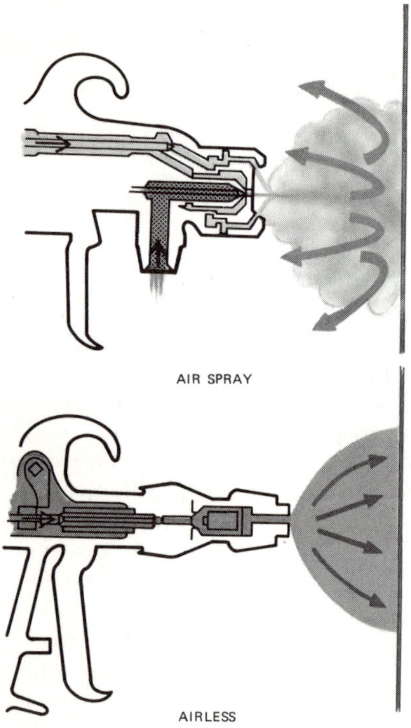

Fig. 3–1. Differences in the spray patterns between airless and air spray equipment. *(Courtesy, Gray Company, Inc.)*

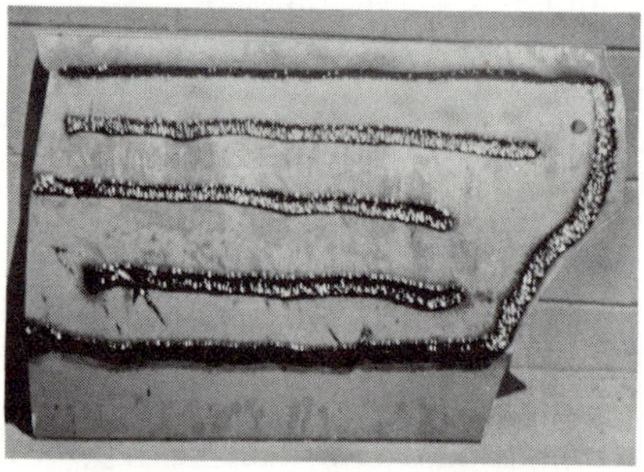

Fig. 3–2. Typical surface after adhesive application (airless) by an automotive manufacturer. *(Courtesy, Gray Company, Inc.)*

SELECTION OF EQUIPMENT

The selection of equipment for dispensing adhesives will depend on which methods are best suited for applying a particular adhesive. Since there are so many adhesives to choose from, the selection of equipment can sometimes be a problem.

If an adhesive and a method of application have been selected, the choice of equipment is still to be decided. Before buying any particular equipment, certain minimum requirements should be expected from the equipment.

1) Equipment freeze-up is bound to occur either from power failure or some other cause. When it does, the equipment should have features that will aid rather than hinder disassembly and cleanup.
2) If the equipment uses air or hydraulic pressure, leakage can be a problem. Sometimes a close regulation of pressure is required, and whether or not this can be done with ease is important to know.
3) The life expectancy of the equipment and components should be known since replacement and overhaul are necessarily part of the future cost of manufacturing. Troubleshooting charts for the equipment are helpful since they aid early detection and quick repair of trouble.
4) Production rates for the equipment should be known as well as whether these rates are for continuous or intermittent production.
5) Means for the detection of trouble are important since the equipment will have to be quickly purged if trouble is detected.
6) Quality control will demand that a convenient way of checking the proportions of the various adhesive components be available to them.
7) The control of adhesive temperature and the use of temperature indicators may be necessary for some adhesives.
8) The equipment should allow for the drawing off of any portion of the adhesive components whenever necessary.
9) The handling of equipment should be as simple as possible if a shop operator is to use it. If complicated, only an engineer, a set-up man, or a maintenance man will be able to master it. Simplicity of application also assures the quick training of new workers. As a general rule, if the equipment and method of application are simple to operate and apply, the reliability will be higher.
10) When equipment breaks down, service backup will be necessary. Availability of a manufacturer's field service personnel, spare parts as well as schematics, and parts lists all help shorten the amount of time the equipment is down.
11) Production objectives should be considered and compared with the capabilities of the proposed equipment. Cleanup, cure out, setup time, maintenance of any equipment all affect production. The speed and consistency of unit production are closely tied to the selection of the right equipment for the job.
12) A delineation of the warranty coverage would be important to anyone thinking of buying equipment. If the warranties are also aided by the availability of installation and operation manuals, wiring diagrams, etc., and other drawings, the chances are greater that the purchase of the new equipment will be a sound one.
13) Extreme care should be taken to make sure that the chemical contents of the adhesive are fully compatible with all of the parts of the equipment with which the adhesive may come in contact. Experience has shown that many types of adhesive can cause corrosion, pitting, and failure of certain types of metal, equipment, plastic parts, packing, etc.

An understanding of the various methods of applying an adhesive are important to consider before ordering any equipment. Each equipment type has its own backup equipment and supplies, its own assembly speeds, and its own safety considerations (see Table III–2).

Table III–2. Adhesive Equipment Application Considerations.

Type	*Equipment Requirements*	*Speed of Assembly*	*Safety Considerations*
Airless Spray	Pump Spray gun Material container Air compressor	High coverage low time High production Potential very suitable for assembly Production line usage High production of units Low labor per unit sprayed	Material source can be covered at all times and under pressure High pressure guns have safety latch Static line from unit ventilation
Air Spray	Pump Spray gun Material container Air compressor	High coverage low time High production Potential very suitable for assembly Production line usage High production of units Low labor per unit sprayed	Material source can be covered at all times and under pressure Pressure guns have safety latch Ventilation Static line from unit
Roller	Material container Roller Cleanup solvent	Coverage for flat piecework But high unit production is seldom obtained Flow roller type reduces the manual labor by pressure feeding the roller	Open material source Drips and dabs form under piecework Ventilation If used, extension device should be firmly attached
Brush	Material container Brush Cleanup solvent	Slow assembly unit by unit basis Low unit quantity generally the least efficient of commonly used methods	Open material source Drips and dabs form under piecework Ventilate for solvents only
Extrusion Flow	Pump Material container Extrusion-flow gun Cleanup solvent	Excellent control of adhesive at the point of actual application Used often where a bead or ribbon is needed	Open material source Remove all air from system before operation Ventilate for solvents only
Trowel Knife Roller	Material container Trowel notched trowel, or knife roller Cleanup solvent	Poor for large part lots but excellent for small areas whose adhesive does not readily spray by airless or air spray Fairly constant amount applied Good for the building trade	Open material source Ventilate for solvents only
Dip Immersion	Dip-immersion reservoir Dip rack Part hooks	Excellent for large lots of uniform size Material is not economically used to its best potential Rack and hooks normally need periodic cleaning	Open material source Ventilate for solvents only

UNIT COST CONSIDERATIONS

Consideration of cost for any unit is necessary if a realistic decision is to be made on whether or not the adhesive process is to replace a given method of fabrication. The following study will not produce an exact cost figure, but it will be capable of producing a high probability of falling within a certain range of accuracy.

Materials

The actual cost of the various components in an adhesive is usually figured per pound or per gallon. This cost should include the following: (1) actual material used for production after determination of the product waste, (2) evaporation loss, (3) excess adhesive application, (4) amount of adhesive remaining in containers, (5) solvent cleanup, and (6) material loss due to contamination.

Size of Container and Type

Bulk Considerations. Variations in material price is determined by container sizes such as 1 gal cans, 5 gal pails, 50 gal drums, or tank car lots. Larger containers usually have a lower fob. price. The selection of the container size is dependent upon the material usage within its shelf life limitations.

In-Plant Container Handling. Storage of material prior to transfer should include a temperature controlled environment and a means of stock rotation (first in, first out). Vehicles must be available for container transfer. Restriction in floor loadings must be considered.

Material Transfer for Production. The method of material removal from the container must be considered such as the use of storage tanks, pumps, and piping.

Special Container Types. Products requiring agitation such as the neoprene adhesives require agitation drums which are more expensive than ordinary drums. Most airless spray adhesives require a pressure pot, while heavy mastics require reinforced drums. In some cases, plastic lined special non-corrosive stainless steel containers are required to prevent a possible chemical change or breakdown in certain types of adhesive materials.

Equipment

The type of adhesive and the production rate of the product help to determine the type of equipment to use and therefore the cost of the equipment necessary. Most solvent and water-based adhesives can be applied by simple spray equipment utilizing either air spray or airless spray. Plant air pressure is a requirement for these applications involving high volume usage. Smaller applications might require a simple hand-gun, a roller, or brushes. The heavier products such as multi-component polysulfides, epoxies, and some urethanes require complex metering equipment for the proper discharge of the varied component materials. Smaller applications for these high viscosity adhesives might require simple hand mixing and could be applied by either a hand extruding gun, a roller, or a brush. The hot melt materials require heating prior to application while most tapes usually require holding fixtures or rollers with which to apply pressure.

Manpower

The labor needed for the application of adhesives is dependent upon the following:

1) The size and type of equipment being used
2) Complexity of the job skill required to operate the equipment
3) Production unit rate per hour, week, etc.
4) Cleanup requirements
5) Finished product inspection.

Additional Safety Requirements

Ventilating Equipment. The use of solvent-based adhesives requires necessary air movement not only for the health and safety of the operator but also for hastening the curing of the adhesive. Where combustible solvents represent a large percentage of the adhesive composition, ventilating equipment is required. Fire and insurance codes are also involved.

Solvent Cans. Some solvents require special containers due to their combustibility and odor. The standard red label cans are required when most ketones are used. Some of the chlorinated solvents are also included in this group.

Protective Clothing and Skin Care. Spray booth operators who are confined must wear protective coveralls and face hoods in order to eliminate the high concentration of solvent. Other applications might require the use of special gloves or hand creams to protect individuals susceptible to allergies. Irritants might include formaldehyde or phenolic resins found in many of the heat reactive adhesives and some of the organic solvents in many of the nitrile rubber spray adhesives. If some adhesives are mixed dry and raise a fine dust, face masks should be used to filter out the dust.

PLANT FACILITIES

Size of Manufacturing Area

Area requirements for different pieces of equipment are obviously a function of the product, the application equipment involved, and the possible need for support equipment such as conveyors, exhaust systems, curing ovens, washing or surface preparation units, and material flush receivers. Therefore an attempt will be made to provide general guidelines based upon some typical applications.

Spray Methods of Adhesive Application to a Small Area. Both airless and air spray methods can be used for applying adhesive.

Airless Spray. Airless spray equipment for applying adhesive to a small area (2 sq ft or smaller) when using a pumping system for a 55 gal drum (see Fig. 3–3) requires from 6 to 10 sq ft of floor area. The pumping system should be located as close as is practical to the work area to avoid loss of pressure in the adhesive circuit. Hoses may be suspended, and the operator will require a minimum of 12 sq ft working area, assuming that parts are conveyed to him. A minimum paint booth 8 ft wide by 4 to 6 ft deep may be required, including operator's work area, depending upon local conditions. It is possible to transfer adhesives under low pressure from any remote free area to a wall-mounted high pressure airless pump and thus move the 6 to 10 sq ft storage area requirement to any suitable location.

Air Spray Method. Air spraying equipment for applying adhesive to a small area (2 sq ft or smaller) when using a drum size pumping system (see Figs. 3–4

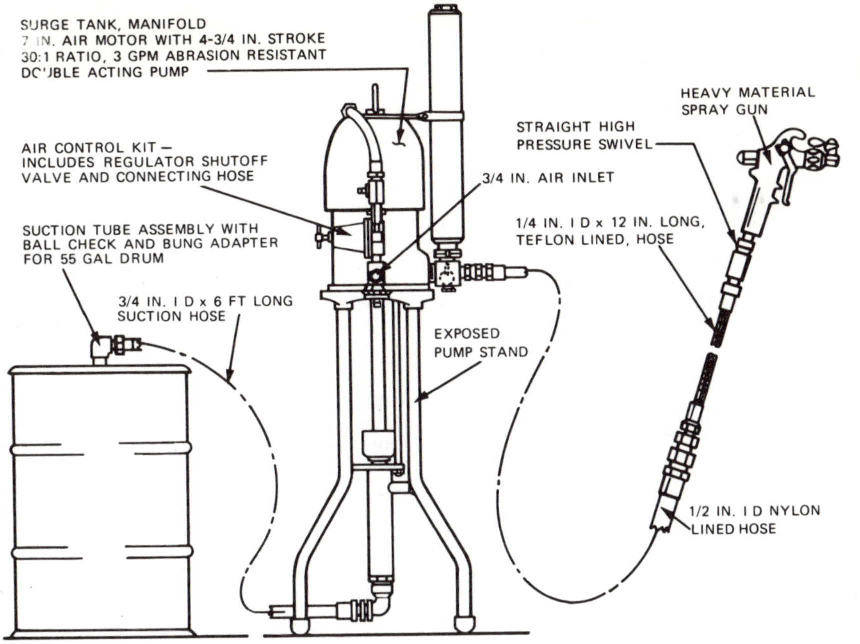

Fig. 3–3. Equipment layout for airless spray installation using a 55 gal drum.

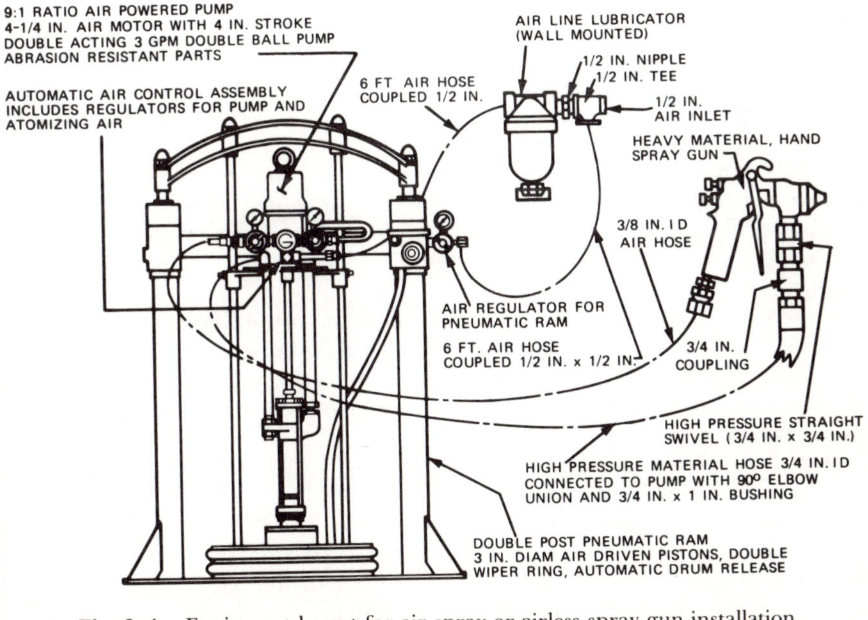

Fig. 3–4. Equipment layout for air spray or airless spray gun installation.

and 3–5) requires the same area as is necessary for the airless method discussed above except for the mandatory addition of a spray booth.

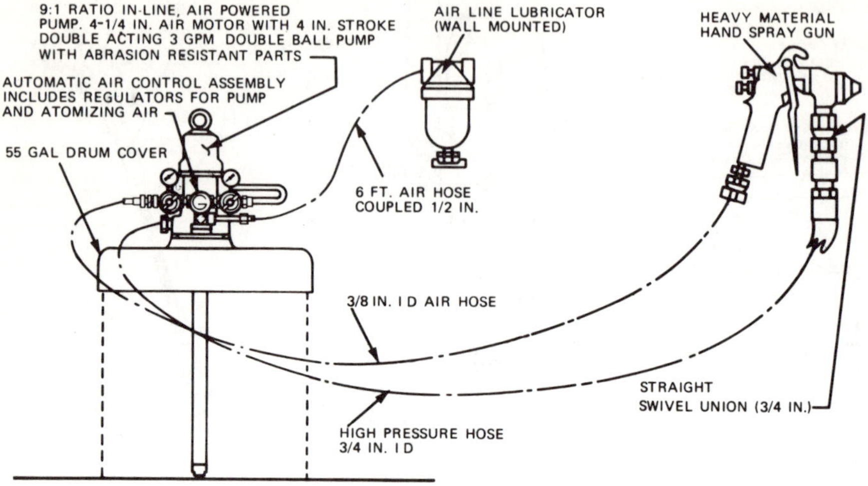

Fig. 3–5. Equipment layout for air spray installation.

Extrusion or Flow of High Viscosity Adhesive. Supply pumps for drum application of high viscosity materials should be located as close to the application point as possible to avoid pressure loss. Pressure loss in long lines is a function of the surface area of the hose and the length. A single gun application from a drum pump requires 6 to 10 sq ft of pump area near the application point. Multiple gun applications are possible along a moving line or series of lines by placing a bank of drum pumps as near the center of all application points as possible and using 2 in. i.d. high pressure (.150 in. wall) seamless steel tubing as an overhead manifold with ½ in. hose drops to the dispensing guns at each application point. Localization of a series of pumps requires about 8 sq ft per pump. Provision must be made at that point for adequate access to replace the empty drum. Commonly used in automotive, truck trailer, and mobile home plants, the central pump manifold system offers several advantages. It reduces the space required to change drums and localizes pumps for maintenance.

Often stand-by pumps are used in such areas with independent valves between each pump and the manifold. This permits shut-off of any given pump for maintenance or drum change without interrupting manifold function. Whistle stop controls are commonly used on such pumps to emit a signal when the drum is empty.

For low volume applications requiring 5 to 15 gal daily of high viscosity adhesive, a 5 gal elevator pump will occupy about 4 sq ft and is usually built on casters for limited mobility. See Fig. 3–6.

This type pump can supply one or more automatic guns which might be mounted in any direct relationship to a moving or interrupted conveyor line to place controlled spots, lines, beads, or ribbons of adhesives to a product.

The system requires from 5 to 15 cfm of air at 50 to 100 psi to apply single component mastic sealers and adhesives. The system is also available in 55 gal drum-size. Similar setups for mid-viscosity (level seeking) adhesives are available without the pump elevator. Low viscosity fluid adhesives may be applied similarly using a pressure pot in lieu of a pump supply. Costs range from $500 to $1500, depending upon size of supply unit and number of application guns.

Two- or Three-Component High Viscosity Extrusion Flow Metering Unit. A typical two- or three-component high viscosity extrusion flow metering unit is used to apply polysulfide to automotive windshields (Fig. 3–7). The space

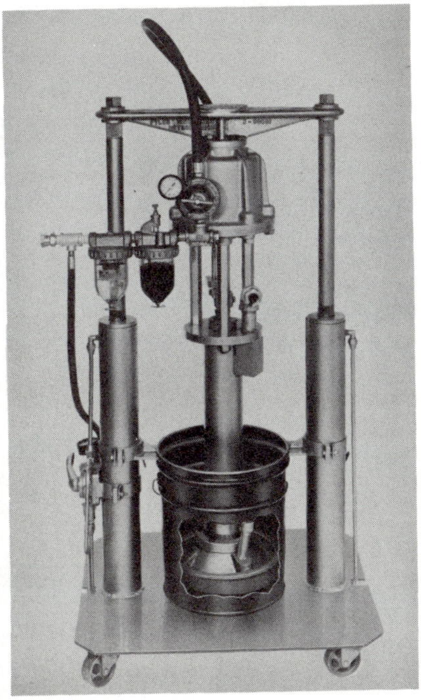

Fig. 3–6. Pump, 5 gal, pail size. *(Courtesy, Pyles Industries, Inc.)*

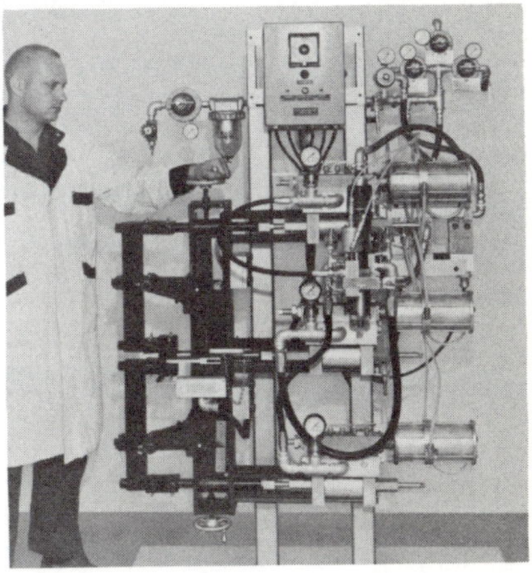

Fig. 3–7. A three component high viscosity meter. *(Courtesy, Pyles Industries, Inc.)*

required will depend upon line speed. An area of at least 60 sq ft will be required for the glass turntable, suspended mixer dispenser, and the operator. Supply pumps and a 3 component meter require about 40 sq ft and must be within 20 ft of the applicator if operating pressures are to be maintained low enough to avoid excess seal and meter deterioration. Dispensing rate is limited to the glass turntable and the operator's capability to make a clean bead on the glass—usually from 1 to 2 lb/min. If the glass requires 8 oz, an application time of 30 sec and handling time of at least 60 sec will dictate the number of machines and floor space required to maintain a given line speed. On critical production lines it is common to include one spare or backup meter coupled to the basic supply pumps to rapidly change over in the event of breakdown. The meter alone requires 10 to 12 sq ft. See Fig. 3–8 for a typical two component applicator, Fig. 3–9 for a three component meter only, and Fig. 3–10 for new fixed ratio water injection polysulfide meter.

Fig. 3-8. Manually triggered two component high viscosity reactive resin meter, mixer, and applicator system. *(Courtesy, Pyles Industries, Inc.)*

The manually triggered two component high viscosity reactive resin meter, mixer, and applicator system illustrated in Fig. 3-7 is available with four different catalyst meters. It provides a 4:1 adjustability with volumetric ratios from 1:1 to 50:1 and can accept viscosities as high as hydraulic pumps can supply. A similar three component unit is typical of those used to apply twin catalyst polysulfide adhesives for bonding automotive glass to metal. Fixed ratio units are built to specific ratio specifications from 1:1 to 100:1.5. These units require an air supply of 30 to 100 cfm at 80 to 100 psi for high viscosity products at rates from 1 to 4 lb/min. Costs range from $6,000 to $14,000, depending upon supply equipment, number of components, selection of mixer, and auxiliary equipment.

Fig. 3-7 illustrates a complete applicator unit with twin catalysts. It includes a meter, base, and automatic flush controls. The unit permits adjustable volumetric ratios from 100:2 to 100:8 of each catalyst.

Fig. 3-10 shows an applicator unit with a high viscosity meter and fixed ratio water injection system for water activated polysulfides. It includes a base pump, meter, automatic fail safe, high and low pressure controls, mixer, 20 ft recirculating hose, an electrically controlled dispenser with whip hose, and a water meter fixed at any specific ratio from 1 to 5 percent. The floor space required for pump and meter is 12 sq ft and has a cost range of from $8,000 to $10,000.

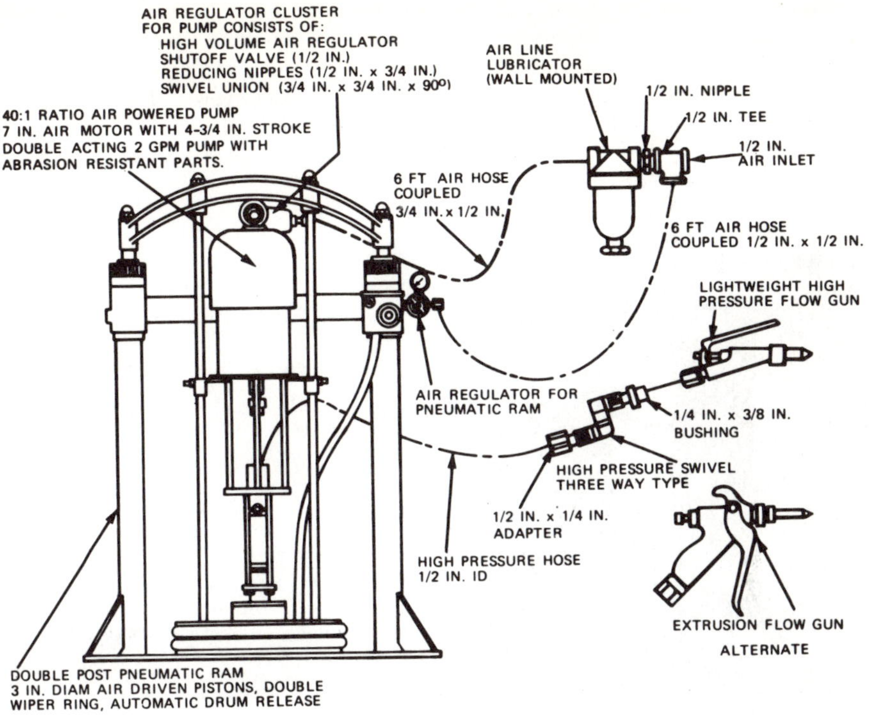

Fig. 3–9. Equipment layout for flow-gun installation.

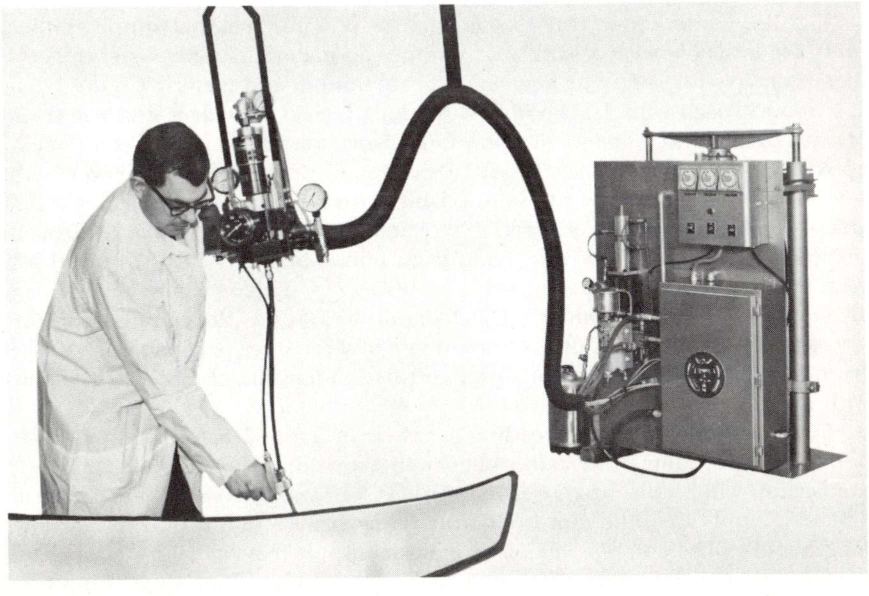

Fig. 3–10. Applicator unit with high viscosity meter and fixed ratio water injection system for water activated polysulfides. *(Courtesy, Pyles Industries, Inc.)*

Extrusion-Shot Machine for Two-Component Low Viscosity Reactive Resin. Fig. 3–11 illustrates a typical extrusion shot machine which requires 10 sq ft of floor space and includes material supply, meter, and mixer. Parts may be fed manually, or automatically using a turntable or conveyor. Two feet must be left behind the machine for replacement of resin containers.

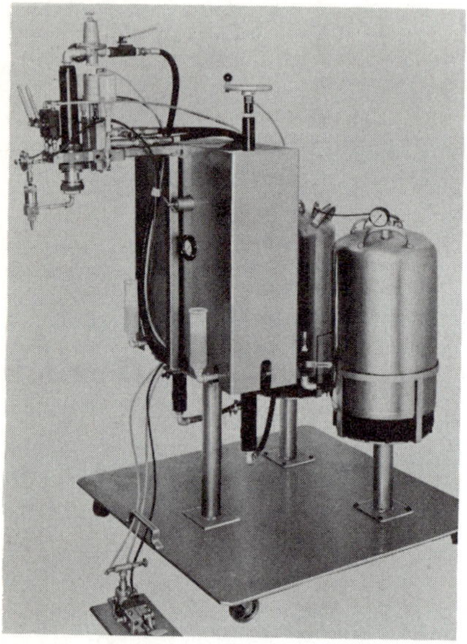

Fig. 3–11. Extrusion-shot machine for use with two component low viscosity reactive resin. *(Courtesy, Pyles Industries, Inc.)*

Fig. 3–11 also shows two 10 gal stainless pots for material supply. Other units are available with a variety of pump type material supply systems, automatic controls for rotary or line conveyor operation, and metering ratios from 1:1 to 50:1, each with 4:1 variability. Various meters provide a shot size from .2 cc to over 100 cc dependent upon ratio. Units are used with two component epoxies, silicone, urethane, polysulfide, or mercaptan-based adhesives with pot life varying from 20 sec to hours. Units with heating adaptations are available to control flowability or reaction rate and are frequently used in controlled bonding of seals to lamp rims, endcaps to filters, fiberglass seats to metal leg retainers, electronic components in assemblies etc. This machine requires an air supply of 20 to 50 cpm of air at from 50 to 100 psi, depending upon the viscosity, shot size, and at an application rate of 2 to 10 sec per shot. Costs range from $4,000 to $6,000. Similar lower cost fixed ratio models are also available with metering ratios from 1:1 to 6:1.

Curtain, Roller, Knife Spreading for Sheet or Sandwich Application. Basic adhesive application by curtain, roller, knife spreading for sheet or sandwich application will require space related to sheet size and line speed. For a typical 4 ft wide applicator, the unit itself will occupy from 3 to 8 ft of conveyor line length and will be 6 to 8 ft wide. The material supply may be from a hopper on the unit but may be pumped from virtually any location. Line lengths must be related to the need for sheet loaders, the applicator, secondary loaders, compression rollers, drying time or oven area. The use of a multiple stack pressure-oven cure must be considered as opposed to a roller pressure oven on line cure and is dependent upon both the product and the adhesive used.

Hand Application Methods. The preceding discussion has mentioned brushes, rollers, and independent caulking guns (see Figs. 3–12 and 3–13). Space requirements related to such equipment are obviously negligible.

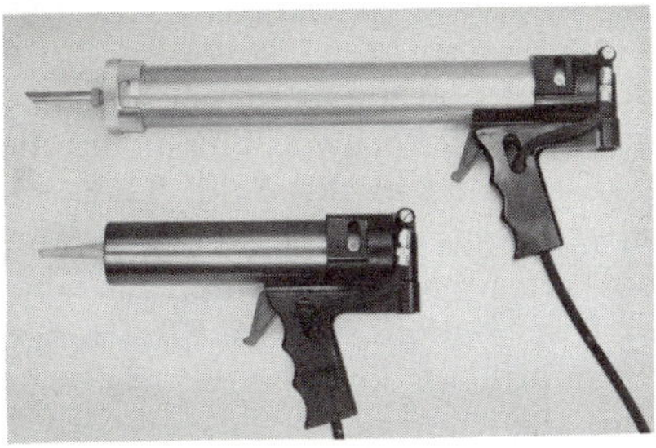

Fig. 3–12. Air operated manual caulking guns. *(Courtesy, Pyles Industries, Inc.)*

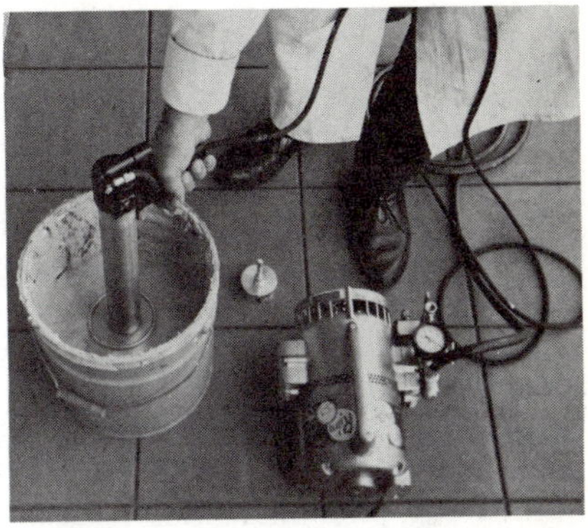

Fig. 3–13. Cartridge caulking gun and compressor. *(Courtesy, Pyles Industries, Inc.)*

Several kits are marketed which include various combinations of the following: cartridge caulking gun, $\frac{1}{4}$ to $\frac{3}{4}$ hp compressors, various hoses and controls, bulk loading barrel with cap and piston, 1 gal pail follower loading plate, 5 gal pail follower loading plate, and an assortment of nozzles. These kits may be used in any adhesive application where high mobility is required and volume does not justify the use of hydraulic drum pumping equipment. All components may be variously combined. For example, a bulk loading gun and vacuum regulator could be used on a plant air supply to vacuum load and dispense adhesives from 1 or 5 gal pails with the appropriate follower loader plate, useful with any cartridge loaded material and pail materials up to several hundred thousand cps viscosity. Such a combination will readily load adhesives having the consistency of putty. Costs vary: a single gun can be under $40; a complete kit can go up to $350.

Some manufacturers have a sample adhesive application kit for mixed reactive resins available. It includes: sealant gun with 6 oz cartridge retainer and air hose, 6 oz cartridges, plungers, nozzles, front caps, rear covers, and a compressed air line connector.

Disposable polyethylene cartridges are usually loaded with reactive resin from an appropriate mixing machine and may be shock frozen at −80°F, stored (subject to material manufacturer's instructions), thawed, and dispensed, or they may be dispensed promptly after filling. The air gun may be operated from any compressed air supply. Guns and cartridges are available in $2\frac{1}{2}$, 6, and 12 oz size with cartridge and retainer as the only variables. Gun costs vary in the $50 area, and plastic parts cost from $1\frac{1}{2}$ cents to 20 cents depending on the quantity purchased. Over 20 varieties of disposable nozzles are available with length from 2½ to 10 in., orifice variations from $^1/_{32}$ in. to $^3/_{16}$ in., and a number of frequently ordered formed-outlets.

Most frequent usage is in the application of polysulfides, epoxies, urethanes, and silicones as adhesive sealants in airframe, electronic, dual insulated glass, and building construction.

Equipment Availability

A definitive treatment of equipment from all manufacturers is impossible, especially since the adhesive industry is in a state of rapid growth, a fact which makes it difficult to know all the equipment available now or in development for the future. The development of new adhesives will create a constant change in the technology of application. The following is a resumé of commonly known equipment, available with wide distribution, and with capable service support.

Airless and Air Spray Equipment. Air operated reciprocating hydraulic pumps are most widely known, used, and distributed in this field. They are available in three basic forms for direct use with 5 gal pails, 55 gal drums, and independently mounted on stands or wall brackets for supply through a flexible "snorkel" tube or by gravity from tanks or from remote sources. Sizes range from about 1½ in. diam 1:1 ratio transfer pumps for low viscosity fluids to pumps with pneumatic motors as large as 8 in. diam and power ratios from 2:1 to 80:1. Ratio in this sense is the power multiplication factor of the air pressure used to drive the pump motor. For example, at 38:1 ratio pump would produce 3800 psi fluid pressure at its outlet with an air input of 100 psi.

A number of manufacturers utilize electrically driven hydraulic power units in highly efficient, portable airless spray units with output pressures in the 3,000 psi range (see Fig. 3–14).

Heated systems are available which permit the use of adhesives whose room temperature viscosity is too high to permit spraying.

Many low viscosity adhesives are supplied to air spray guns from inexpensive

Fig. 3–14. Airless pump used for spray application. *(Courtesy, Nordson Corporation)*

pressure pots, and they are available from pint size to 55 gal size and at least one manufacturer produces glass-lined, quick opening pots for ease of cleanup.

All spray equipment mentioned is available virtually off-the-shelf in most industrial areas.

Extrusion Equipment for High Viscosity Adhesives (See Fig. 3–15). As with sprayed adhesives, the pneumatic reciprocating pump is the predominate unit for moving and applying extruded adhesives. Pump characteristics are similar

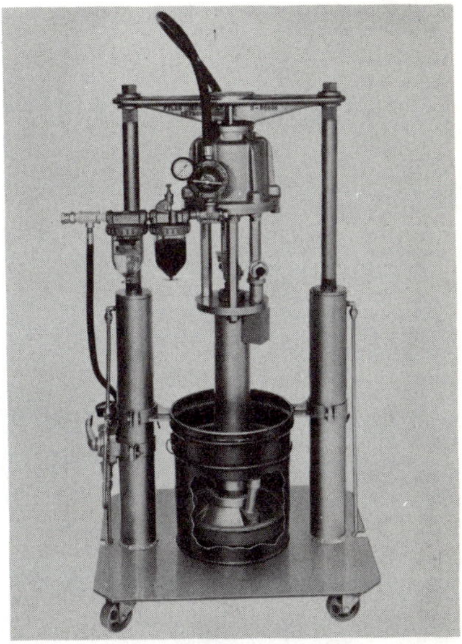

Fig. 3–15. Pumping equipment. *(Courtesy, Pyles Industries, Inc.)*

to those used with spray materials with a few notable exceptions. Most require the use of a follower or pressure plate fixed to the bottom of the pump which forces the mastic into the pump and wipes the sides of the container. The pump is then mounted on a ram or elevator which guides it into the container and in most cases provides up and down pressure through one or two post type pistons and cylinders. Units are available with motor sizes from 3 in. diam to 8 in. diam and ratios from 9:1 to 80:1 for use with 5 gal, 120 lb, and 55 gal pails or drums.

Pumps of this type for 5 gal pails are often stocked locally by distributors, but the drum size units because of their bulk are usually stocked at the point of manufacture or in a manufacturer's local branch warehouse. In most areas they should be available in from one to three weeks.

A limited number of high viscosity adhesives are available which cannot be pumped. In some instances, jacket pail or drum warmers will reduce their viscosity sufficiently to permit pumping. If not, the use of screw type extruders is necessary.

These adhesives are normally applied with a wide variety of available guns. All involve a valving mechanism to shut off flow either at the end of the gun where nozzles are attached or at the tip of specially prepared nozzles. They are available in pistol grip (see Fig. 3–16) or palm grip form and in a number of air operated piston forms for automatic applications.

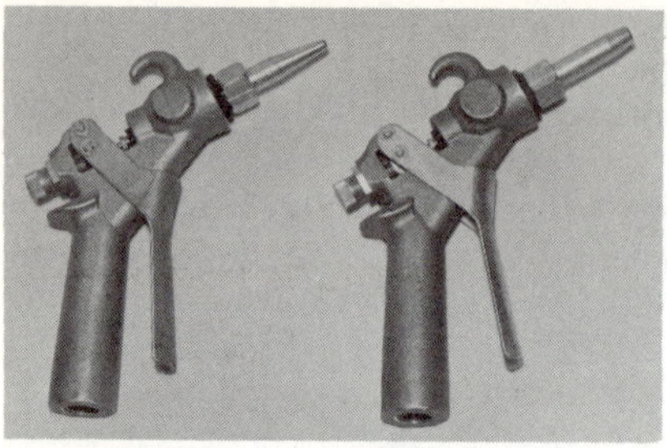

Fig. 3–16. Palm grip nozzles. *(Courtesy, Pyles Industries, Inc.)*

Generally the gun is used to stall the pump, but accessory valving kits are available which shut off the air to the pump when the trigger is released.

Dispensing nozzles are available in infinite variety (see Fig. 3–17) as are attachments for the guns which incorporate either flow through brushes or side mounted brushes to spread the adhesive. Most local distributors stock the manual type guns.

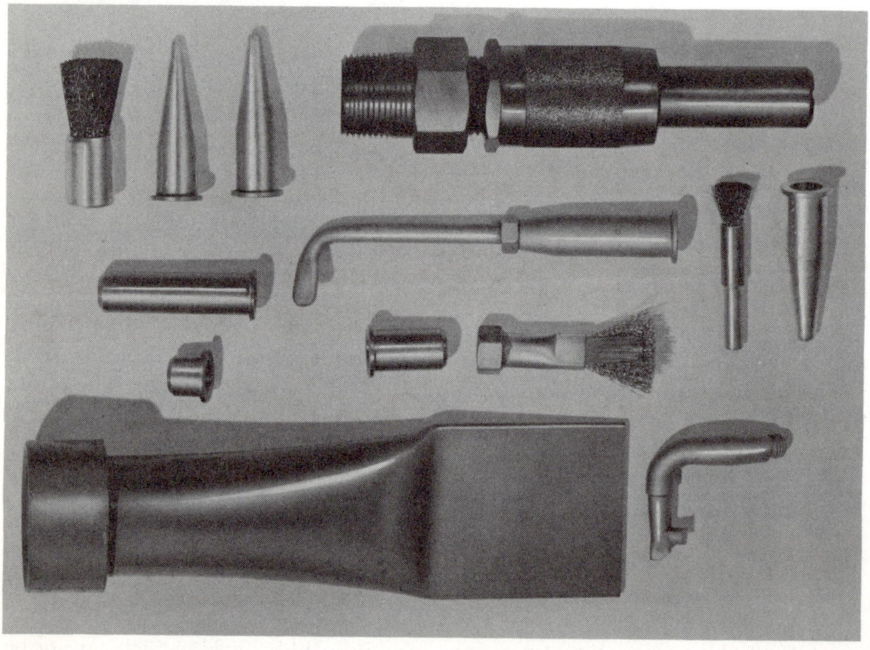

Fig. 3–17. Dispensing nozzles. *(Courtesy, Pyles Industries, Inc.)*

Special measuring dispensers are also commonly available. They use either material pressure (see Fig. 3–18) or auxiliary air pressure through a piston to eject a measured repeatable shot of adhesive. Units of this type are often mounted in multiples on a fixture to apply spots of adhesive to such products as the spider reinforcement on automotive hoods or trunk compartment lids.

Extrusion Equipment for High Viscosity Multiple Component Reactive Resins. Equipment of this type generally consists of reciprocating pumps which

Fig. 3–18. Pressure regulator. *(Courtesy, Pyles Industries, Inc.)*

supply components under pressures of 500 to 2,000 psi to meters which are linked together to provide simultaneous supply of the two or three components in their proper proportions to a mixing head. The meters may be rigidly linked end to end (see Fig. 3–19) in which case the material proportions are fixed and

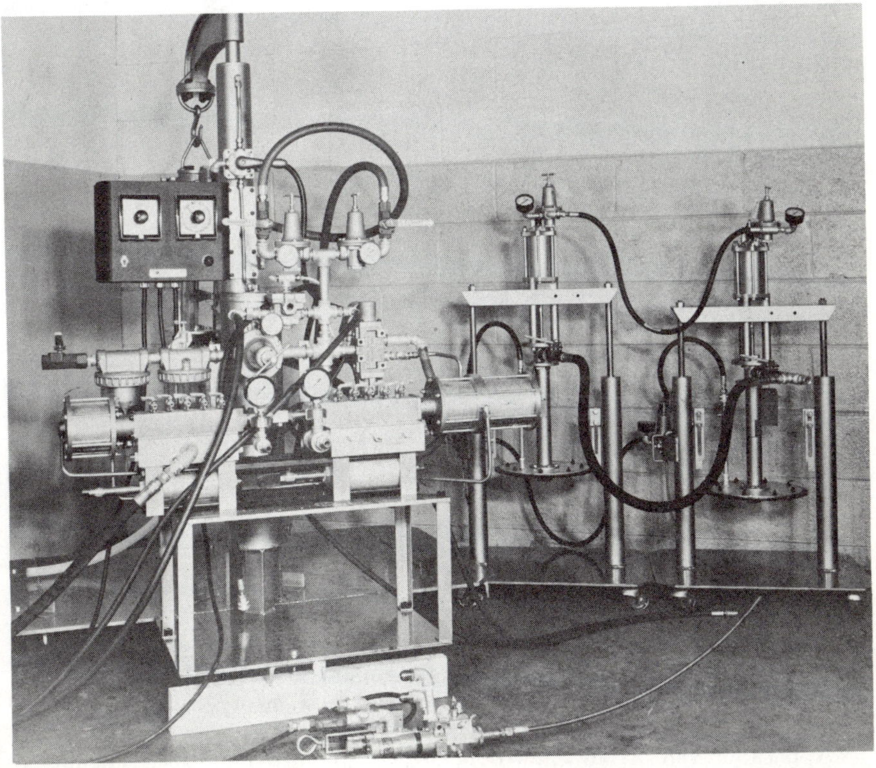

Fig. 3–19. Extrusion equipment for high viscosity multiple component reactive resins. *(Courtesy, Pyles Industries, Inc.)*

the catalyst or hardener cylinder and piston are specially manufactured for the material to be used.

Similar meters may be linked through lever arms so that the stroke length of one or more meters may be varied to permit adjustment of the ingredients (see Figs. 3–20, 3–21, and 3–22).

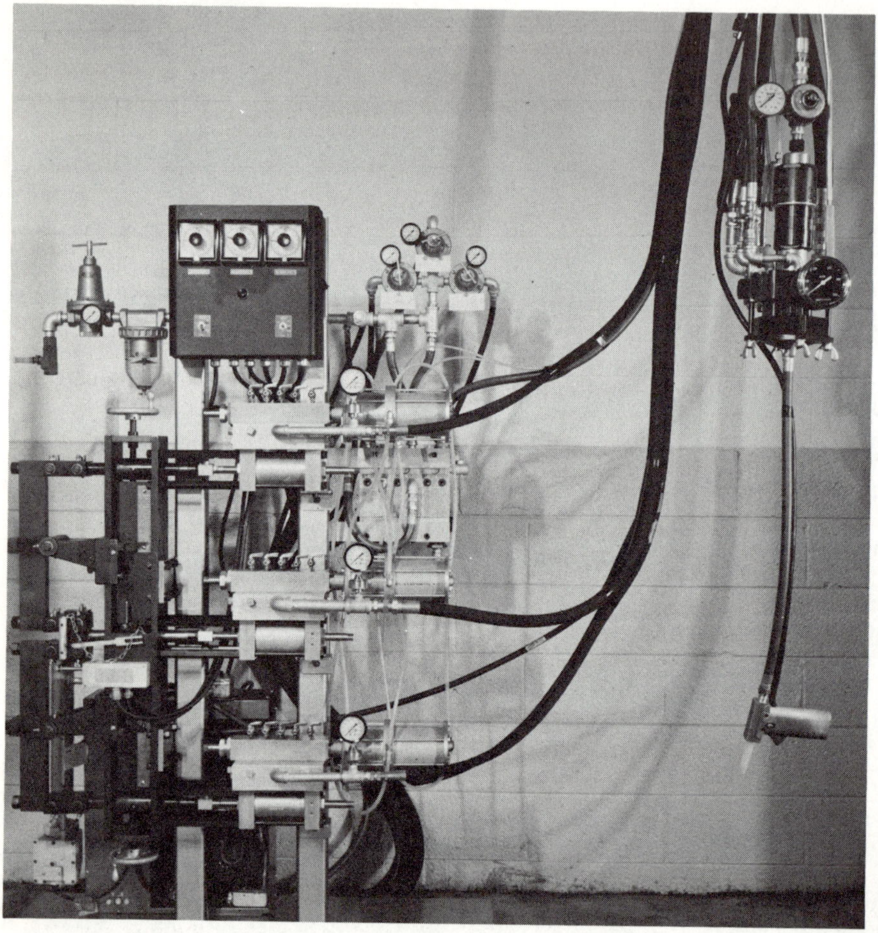

Fig. 3–20. Metering equipment. *(Courtesy, Pyles Industries, Inc.)*

Mixing devices vary, depending upon material characteristics, viscosity, pot life, and the required flow rate. Virtually all are driven by an air, hydraulic, or electric motor. The air driven devices are preferred since they are stallable in the event of an inadvertent cure in the mixing chamber. These units must all be readily disassembled for cleaning and may incorporate solvent flush attachments.

Application rates from a few ounces to 20 lb/min are possible. The meter-mix machines remain a somewhat customized product, although substantial

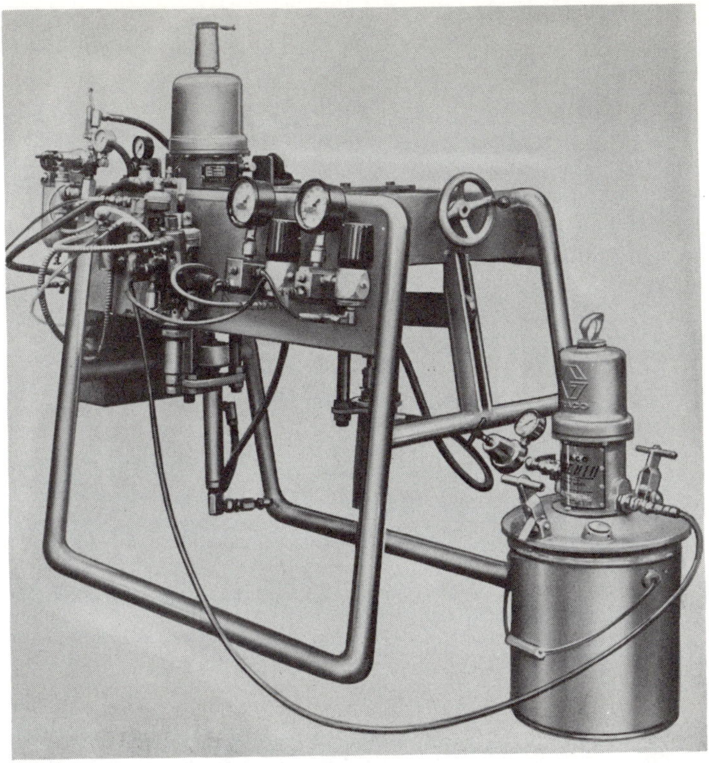

Fig. 3–21. Variable ratio with flush system and adjustable stroke mechanism on primary pump. *(Courtesy, Gray Company, Inc.)*

standardization has been made. They are available from a very limited number of manufacturers but are usually unavailable on a local basis. They require 4 to 16 weeks to procure, depending upon the complexity of the customized features.

Customizing and accessories are available in the form of counting or timing dispense controls, special application heads or manifolds, automatic dispensing cycles, automatic material purge, automatic shutdown in the event of loss of a component through malfunction or depletion of supply, and a host of others.

Dispensing Equipment for Low to Medium Viscosity Reactive Multiple Component Resins. Much of the discussion on high viscosity resins applies also to dispensing equipment for low to medium viscosity reactive multiple component resins with the obvious exception that flowable or level seeking liquids usually do not require independent supply pumps since the supply pump itself is designed as a meter. Coupling such pumps rigidly or through an adjustable linkage in which two or more are driven by a single air motor permits metering the components from their original containers. Positive displacement reciprocating meters of this type provide accurate metering, without regard for fluctuations in viscosity, but heat may be applied to any adhesive component to improve its flowability or to influence reaction conditions.

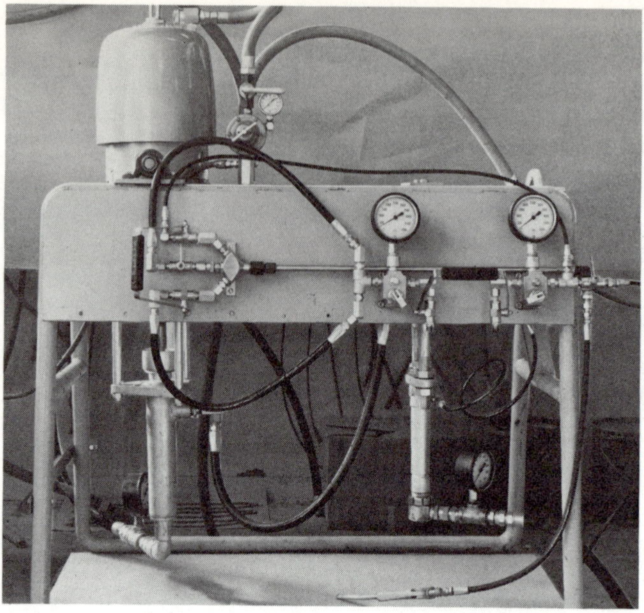

Fig. 3–22. Variable ratio extrusion equipment for polysulfides. *(Courtesy, Gray Company, Inc.)*

Other common metering equipment for reactive fluids use gear pumps which are either linked to a common drive source or are driven independently by synchronous motors. The latter may be immersed in the resin supply to eliminate problems attendant upon exposure of shaft seals to air and moisture which tend to react with many resin or catalyst components. Units of this type must operate with a controlled viscosity to meter accurately, necessitating that materials must be consistently and constantly recirculated and controlled in temperature. Such units are readily available from small hand crank models (see Fig. 3–23) which simply dispense two metered streams for hand mixing, to units which meter 5 or more independent components at rates over 100 lb/min.

Fig. 3–23. Hand crank gear pump for dispensing two metered streams. *(Courtesy, Pyles Industries, Inc.)*

A third basic type equipment in this area incorporates the essentials of the linked pumps with an adjustable stroke drive motor (see Figs. 3–24 and 3–25). Adhesive volume for any application is controlled by equipment that dispenses repeatable shots of exact quantity.

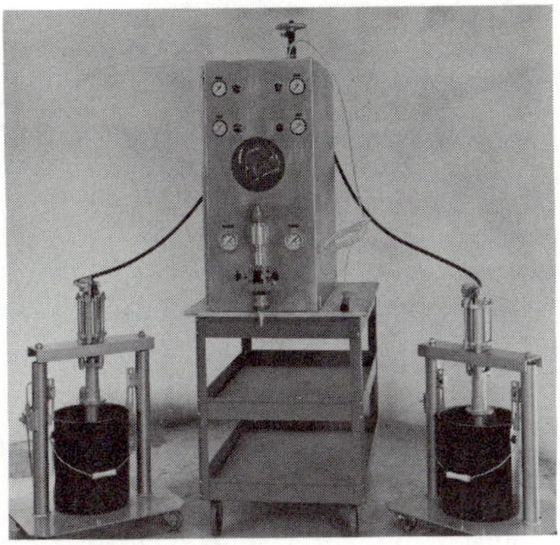

Fig. 3–24. Pump with adjustable stroke drive motor. *(Courtesy, Pyles Industries, Inc.)*

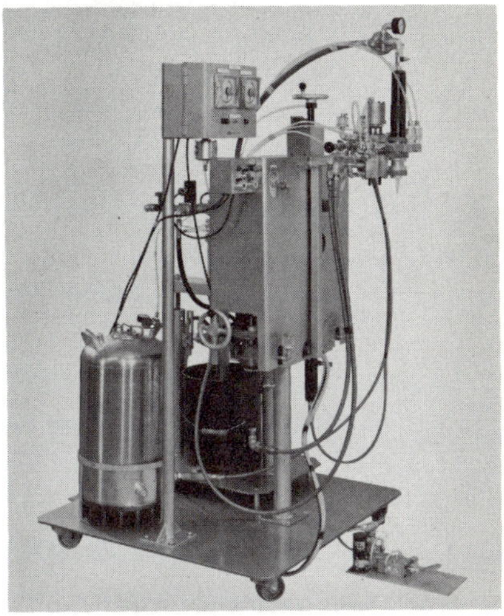

Fig. 3–25. Pump with adjustable stroke drive motor. *(Courtesy, Pyles Industries, Inc.)*

Again the number of manufacturers for equipment capable of handling low viscosity reactive resin is small; customizing is frequently required, and delivery might well require 4 weeks to 6 months, depending upon the degree of specialization required.

Curtain, Blade, and Roller Coaters. Adhesive applicators of this type are generally associated with more or less continuous run conveyorized lines in which sheets, plates, coils, or rolls of strata are coated with adhesive at high speed (see Fig. 3–26). Curtain and roll coaters are generally used with low viscosity adhesives or adhesives in which viscosity is reduced substantially by heating the material supply, applicators, and even the substrate and environment. Blade and roller coaters may be used with higher viscosities up to thixotropic mastics.

Roll coaters are often used for the application of adhesives to two sides of a material which will be laminated. Multiple sheets and complex feed arrangements with multiple roll coaters can produce multilaminar products. Most require auxiliary equipment such as strata feeders, conveyor lines, compression rollers, cutoff devices, pump or hopper adhesive feed, and ovens. Line operating speeds of hundreds of feet per minute are possible.

Although based on relatively standard components, such units are usually customized to a degree, and availability in most areas is limited to the manufacturer's facility. Planning a laminating line of any complexity should be based on 4 to 12 months availability.

Fig. 3–26. Top and bottom glue spreader with water cooled doctor rolls. *(Courtesy, Black Brothers Company, Inc.)*

Hot Melt Adhesive Applicators. Applicator units for hot melt adhesives take four general forms. First, the oldest and most widely known are those which use a formed rod, or strip of adhesive as a piston which is pushed through a heating element, confined in a heated manifold and dispensed through a heated nozzle or by a heated applicator disc which rotates through a molten bath of adhesive. The simplest applicator of this type is available in virtually all hardware stores with sticks of adhesive which are pushed through a gun with the operator's thumb. Complex units can be used to apply solid lines or interrupted dots of adhesive from continuous rolls of formed hot melt at rates of hundreds of feet per minute. These units generally require some customizing which might as a result require a wait of 3 to 6 months.

The second group of hot melt applicators use a melt tank to bring the adhesive to usable viscosity and dispense it either by pneumatic pressure or by pumping. Small units are readily available in which the molten material is withdrawn from the tank with a dispenser much like a grease gun. More complex units provide recirculation of the adhesive when not dispensing and are capable of dot, bead, or spray application at high rates of speed (see Figs. 3–27 and 3–28). One to three months should be considered normal delivery for these units, but some standard models for basic carton sealing may be locally available in major industrial areas. These units are highly satisfactory if the adhesive material does not degrade from being held at application temperature in bulk.

A third category is in the test marketing stage now with only limited application data available. It consists of conventional mastic hydraulic pumps on elevators with heated follower plates, pumps, conduits, and dispensing guns. These pumps heat only the surface material in a pail or drum, thus avoiding holding the bulk of material at operating temperatures. They require fairly complex heating elements, unique insulation, and sophisticated heating and sensing controls.

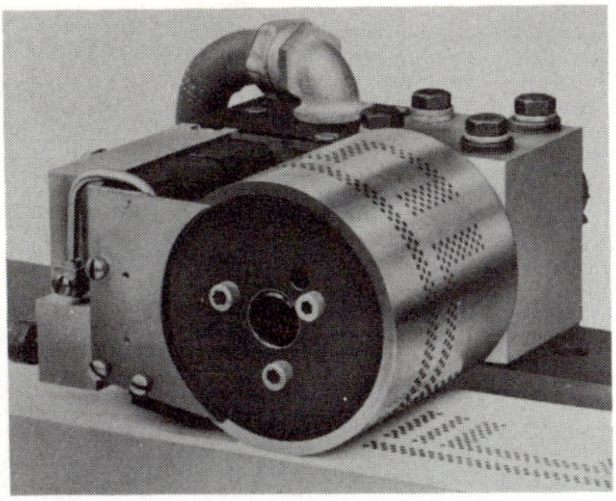

Fig. 3–27. Hot melt adhesive wheel applicator. *(Courtesy, Nordson Corporation)*

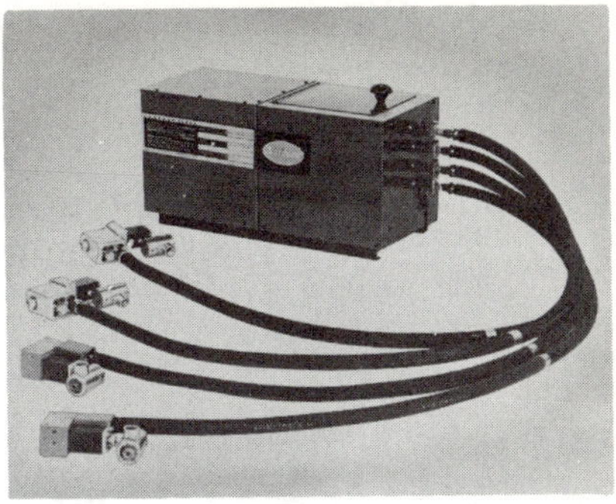

Fig. 3–28. Hot melt applicator. *(Courtesy, Nordson Corporation)*

Specialty Glue Applicators. Unique but standard applicators for low to medium viscosity adhesives which range from dowel hole dispensers (see Fig. 3–29) to multiple dot or bead applicators (see Fig. 3–30) are available in virtually every industrial area. They are widely used in the furniture, paper, and carton sealing industry (see Fig. 3–31).

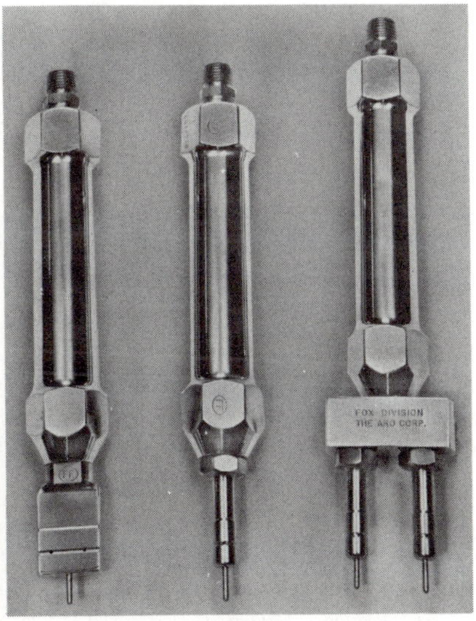

Fig. 3–29. Dowel hole dispensers, left to right: mortise, dowel, and dual doweler. *(Courtesy, Fox Division)*

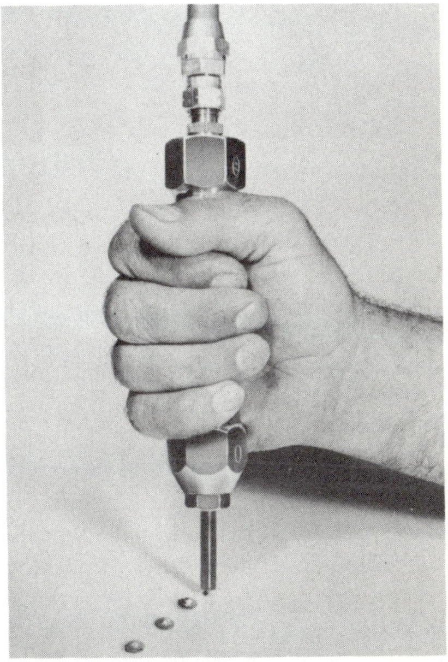

Fig. 3–30. Spot gun. *(Courtesy, Fox Division)*

Hand Applicators

Brushes, squeeze bottles, tubes, paint rollers, and manual caulking guns are available in most ha dware stores. Air operated caulking guns of two types are widely available fro n specialty distributors in three basic forms:

1) Air guns which accept standard $\frac{1}{10}$ gal paper, foil, or plastic cartridges of adhesive
2) Similar guns with interchangeable heads to permit bulk loading from 1 or 5 gal pails
3) Air guns which accept throwaway plastic cartridges of $2\frac{1}{2}$, 6, and 12 oz size which may be filled by the user. This type permits attachment of a wide variety of plastic nozzles from $2\frac{1}{2}$ to 10 in. long and with orifices from $\frac{1}{32}$ in. diam to $\frac{1}{8}$ in. $\times$ $\frac{3}{4}$ in. bead.

Fig. 3-31. Case sealer. *(Courtesy, Fox Division)*

Often the best guide to equipment availability may be obtained from the manufacturer of the adhesive under consideration. Many adhesive manufacturers have equipment in their product labs to test materials and to demonstrate their application. In addition, the larger equipment manufacturers maintain demonstration labs where the materials may be tested on typical equipment if necessary. Unbiased information can be obtained from trade associations.

Waste Disposal

In general, there is little or no waste disposal problem with single component compounds beyond normal paint booth maintenance and the care which must be exercised in disposing of degraded solvent based materials with flammable ingredients. High volume applicators such as curtain or roller coaters usually recirculate unused materials to the supply hopper.

With multicomponent reactive resin materials however, unusual provisions must be made. These materials have a pot life, that is, a limited period of time after mixing during which they must be used or they will cure in the mixer and dispenser. Two methods are used to avoid this problem, and each leads to a waste accumulation and elimination problem.

Automatic cycling of equipment may be built in with a series of timers which (1) automatically run the unit if it is not run in production within the allowable pot life period, and (2) run the equipment long enough to fill the mixer dispenser with freshly mixed material. For example, this type of cycling equipment is used with the equipment on automotive glass lines. The dispenser must be located over a disposable container when not in use.

One precaution must be considered in accumulating such waste. Certain epoxies, polyesters, and other compounds may create considerable exothermic heat in reaction, and the effect may be multiplied when using bulk quantities. It is entirely possible that accumulation of a gallon or more of reacting resins with high exotherm can burn spontaneously.

With this cycling equipment, care should be exercised to dispose of the waste material. Containers should be frequently discarded in a closed, fireproof metal waste container.

During prolonged shutdown or in cases in which pot life is below one minute, it may be necessary to purge the mixer-dispenser with a solvent. The supply unit is shut off, and the mixer is operated while a suitable solvent is pumped or pressure fed through the mixer. The unit is then blown clear of solvent with air. When possible a nonflammable chlorinated hydrocarbon-type solvent should be used in a disposable container. Should material characteristics require a flammable solvent such as acetone or alcohol, care must be exercised in disposal as would be applied with any flammable cleaning fluid or lubricant.

In general, local requirements will dictate ultimate waste disposal. It is *never* satisfactory to run them into municipal or private sewerage systems. Burning when permissible under controlled conditions is ideal. In large usage installations, a sump is sometimes used to collect wastes, and liquid residues are pumped out by commerical disposal trucks. Solid or dried waste residue in disposable containers may be disposed of with all other combustible rubbish.

chapter 4

surface preparation for bonding

Surface preparation is one of the most important steps in the entire bonding process. The preparation of the adherend surface for adhesive bonding will largely determine the success or failure of the bond. Consequently, an improperly prepared adherend surface will produce a poor bond.

The purpose of preparing the adherend's surface is to guarantee that after the adhesive is applied, it will actually wet the surface of the adherend and not what appears to be the surface. The apparent surface could be a layer of oil, grease, film, or some other contaminant.

No adherend surface can remain exposed to the atmosphere without accumulating some or even all of these contaminants. Water, either adsorbed or as a free film, is usually present in at least a monolayer thickness. Therefore, the immediate bonding of prepared surfaces minimizes contamination and should be emphasized.

When a delay in bonding a prepared surface is necessary for the purpose of accumulating and processing numerous detailed parts of an assembly, the prepared surfaces should be protected against contamination. The surfaces may be covered with clean paper, cheesecloth, or films such as polyethylene. Moisture and oil from workers' hands can readily contaminate the prepared surface. Therefore, handlers should wear clean gloves and change them frequently.

Surface preparation is the cleaning or treating of metallic or non-metallic surfaces to insure the best possible bond between the surface to be bonded and another substrate. There are many degrees and methods of surface preparation. The degrees can vary or range from the simple dusting or blowing away of loose dirt to the complete removal of all materials which are foreign to the surface itself. Some methods may include the use of hand tools, chemical cleaning, power tools, steam cleaning, or abrasive blast cleaning.

SURFACE PREPARATION FOR METALS

As mentioned earlier under general conditions applying to both metallic and non-metallic surface preparation, metallic surfaces contain contaminations such as dust, dirt, grease, oil, and, in addition, contain oxide resulting from air corrosion. Again, grease or oil should be removed from the surface not only to obtain a bondable surface but also to permit effective treatment of the surface for much improved bondability. Grease and oil do not permit effective chemical etching and must be removed.

The need for chemical etching of a metallic surface vs abrasion or solvent cleaning only is dependent upon the numerous considerations which are basic to the bond strength required, coefficients of thermal expansion of the materials being bonded, the adhesive characteristics (wettability), and other possible considerations.

The adhesive forces at the interface of the adhesive in a substrate are greater than the cohesive forces of the weaker material and also less than the cohesive forces of the stronger material in a properly bonded composite. Failure of the bond occurs in the cohesive forces of the weaker material. In the comparison with a similar composite but with a layer of oil on the surface, failure occurs in the cohesive forces of the oil layer since the cohesive forces of the oil are far lower than either the adhesive or substrate cohesive forces.

SURFACE PREPARATION FOR NONMETALLIC MATERIALS

The fundamentals of surface preparation for nonmetallic materials can be listed as follows:

1) Clean the adherend surface with solvent
2) Abrade or etch
3) Clean with solvent again
4) Handle the prepared surfaces only with clean gloves
5) Employ a protective covering to guard against airborne contaminants
6) Bond as soon as possible after surface preparation to insure optimum adhesive bonding properties.

An important fact to remember when selecting the degree and method of surface preparation is that the service and life expectancy of the bond is directly proportional to the degree of surface preparation, assuming that the adhesive used will withstand the service conditions to which the bonded area is to be exposed.

The recent technological advances in high quality coatings and adhesives have considerably broadened the responsibility of proper surface preparation. Because of the technical nature of these adhesive coatings, the user may have to rely on the manufacturer or distributor of coatings and adhesives for definite recommendations and specifications, both in the degree and method of surface preparation required.

There are two principal methods of preparing a clean adherend surface: 1) abrasive cleaning, and 2) chemical and/or solvent cleaning. They both depend on physical removal of material from the adherend surface, leaving a relatively clean adherend exposed.

ABRASIVE CLEANING

Abrasive cleaning methods include sanding by hand or by power using sandpaper or emery cloth, grit blasting, and tumbling. Vapor honing is not generally used in preparing nonmetallic surfaces because it usually employs water or steam in addition to an abrasive, which may contaminate or degrade non-metals. In all of the abrasive methods mentioned, a hard sharp material is brought into contact with the adherend surface and causes part of that surface to be ground away, exposing clean pure material.

Prior to any abrasive or chemical preparation of the surface, the surface should be cleaned thoroughly with a solvent. Some molded surfaces are coated with an agent to aid removal from a mold during fabrication. Surfaces may be contaminated by oil during a machining operation. These contaminants, if not removed beforehand, may be driven further into the adherend surface during abrasion; or in the case of chemical etching, grease or oil would prevent the etching solution from reacting with the surface of the adherends.

Care must be taken to prevent the abrasive methods from changing the shape or form of the adherend unduly. For example, grit blasting is a powerful process, fully capable of distorting metal plates 50 mils thick.

Abrasive Process

A surface finished with various grades of sandpaper differs from a blasted surface. Grit blasting generally results in radius peaks and valleys with the depth of the valleys dependent on the variables of grit size employed, pressure, and the material being blasted. Sanding results in sharper peaks and valleys than grit blasting. Usually a coarser paper produces deeper peaks and valleys. Abrasion increases the surface area. Each material may require a specific surface finish, and a particular adhesive being used may readily influence this surface finish. Adhesive viscosity largely determines the surface wetting characteristics. A relatively soft nonmetal should not be abraded with a coarse abrasive in attempting to increase the effective bonding area. The "soft" material could fail cohesively at the sharp peaks and valleys at a value considerably below the adhesive bond strength.

The abrasive process results in an accumulation of debris on the surface of the adherend which is being treated. This debris consists of particles of adherend, particles of surface contaminants, and particles of the abrasive media. Naturally, every single particle of this debris must be removed before bonding. This is thoroughly cleaned after the abrasive step has been completed. Any debris on the surface should be brushed off with a clean brush. Cleaning with filtered compressed air, vacuum cleaning, or wiping with clean dry cloths are methods of debris removal which may be used.

The surface of the adherend can be solvent cleaned just prior to bonding. In the case of a porous material, 20 to 30 minutes or more should be allowed following solvent wiping to permit the solvent in the pores of the material to evaporate. Clean, oil-free cloths, or cheesecloth should not be saturated with the solvent, especially when cleaning a porous surface. A saturated cloth does not lift the debris from the surface of the adherend. Excess solvent will also penetrate the pores of the surface. One technique involves wiping the surface with a solvent moistened cloth followed immediately by a second dry clean cloth. This technique is used for nonstructural bonds.

For fiberglass resin-impregnated laminates and reinforced plastics, removing the surface contaminants by abrasion after solvent cleaning provides a good bonding surface. Solvent cleaning again follows the abrasive step after the debris is removed from the surface.

Nonreinforced-type plastics are generally solvent cleaned followed by sanding with 180 to 320 grit sandpaper, depending on the gage of the adherend. A "soft" material will be more susceptible to damage or erosion of material by coarse abrasives; therefore, a finer grit such as No. 320 should be used. Again, solvent wiping follows the abrasive process.

Fiberglass-reinforced plastics can be cleaned by grit blasting, or they can be cleaned with a solvent (trichloroethylene) followed by sanding with 80 to 120 grit and followed by solvent cleaning with trichloroethylene.

Nonreinforced plastics can be cleaned with solvent (solvent should have very little solvating effect on the surface for prevention of distortion or glazing of the plastic surface). The adherend can then be sanded with 180 to 320 grit followed by another solvent cleaning.

In the preparation of ceramic surfaces the porosity will determine the need for abrading the surface. The surface can be solvent cleaned followed by forced drying to remove all traces of the solvent.

The following abrasive method is used for preparing the surface of mild steel or ferrous type steel:

1) Degrease
2) Grit blast or scrub with silicon carbide abrasive and wipe with a solvent such as trichloroethylene
3) Solvent rinse
4) Bond immediately or prime.

NOTE: Application of water to this type of steel results in rapid oxidation not conducive to adhesive bonding.

Abrasive Blast Cleaning

It is generally acknowledged that there is no faster, more thorough, nor more economical method of surface preparation than abrasive blast cleaning. Abrasive blast cleaning is used in many cases to provide varying degrees of surface roughness necessary for some types of surface preparation.

Many different types of abrasives are in common use. The most widely used are silica, sand, and mineral slag by-products. Others include garnet, crushed flint, walnut shells, metal abrasives such as chilled iron or steel shot and grit. A more detailed description of abrasives appears later in this section. Generally, two methods are used to propel the abrasive: (1) centrifugal wheel, and (2) compressed air. The compressed air method is more commonly called sandblasting, although abrasives other than sand are commonly used. Sandblasting is the most widely used method because of its conveniences.

Centrifugal Wheel Method

Electrically driven centrifugal wheels physically throw the abrasive against the surface to be cleaned, and the objects to be cleaned are either placed in front of the wheels or passed by the wheels on a conveyor system. Therefore, this type is found only in permanent installations. Only metallic abrasives or grit are used since this type of machine recovers and reuses the abrasive.

Sandblasting Method

Sandblasting uses compressed air to propel the abrasive particles against the surface to be cleaned. The portability, speed, and flexibility of sandblasting equipment make it ideally suited for all field operations and most shop or industrial operations as well.

Several degrees of surface preparation can be obtained by sandblasting. The user of the adhesive should find out what degree of surface preparation is required and expected.

The four standard specifications for grades of cleaning by blasting are as follows:

1) Brush-off blast cleaning
2) Commerical blast cleaning
3) Near white blast cleaning
4) White metal blast cleaning.

Not only is the degree of cleanliness critically important, but the anchor pattern must be also considered. The anchor pattern is the maximum height of the profile produced on a surface that will not be detrimental to the life of the coating or to the life of the adhesive bonding the adherend.

The surface roughness or the maximum height of the profile of the surface will depend on the maximum permissible size of the abrasive particles being used. The allowable maximum height is dependent, in turn, upon the adhesive thickness to be applied. This maximum profile height is the measurement from the bottoms of the lowest pitch or valley to the top of the highest peak in the anchor pattern (2).

Table IV–1 gives the measurement of the maximum height of the profile produced by a number of different abrasives and actual blast operations.

Table IV–1. Maximum Profile Height for Various Abrasives (3).*

Abrasive	*Maximum Particle Size*	*Maximum Height of Profile (mils)*
Sand, Very fine (Ottawa Silica Sand)	Through 50 mesh†	1.5
Sand, Fine (Ottawa Silica Sand)	Through 40 mesh	1.9
Sand, Medium (Ottawa Silica Sand)	Through 18 mesh	2.5
Sand, Large (River Sand)	Through 12 mesh	2.8
Iron Grit No. G-50‡	Through 25 mesh	3.3
Iron Grit No. G-40	Through 18 mesh	3.6
Iron Grit No. G-25	Through 16 mesh	4
Iron Grit No. G-16	Through 12 mesh	8
Iron Shot No. S-230§	Through 18 mesh	30
Iron Shot No. S-330	Through 16 mesh	3.3
Iron Shot No. S-390	Through 14 mesh	3.6

*Test criteria: Published in 1954 and expanded and reprinted in 1963. Pressure at machine: 80 psi, .516 in. orifice, nozzle length unknown.
†U. S. Sieve Series.
‡Crushed iron grit.
§Operating mixture.

Wet Blasting

When air, an abrasive, and a liquid are used together, the method is called wet blasting. Fine abrasives in the 100 and finer mesh screens can be used in liquid blasting equipment to produce finishes approaching a polish. Such finishes are necessary for equipment and machinery having critical dimensional tolerances.

An advantage of wet blasting is its ability to control dust when sand is used as an abrasive. Sand blasting always creates a certain amount of dust depending upon the type of abrasive used, but when using silica sand to blast concrete or masonry structures, dust is often a major problem. Wet blasting is the easiest way to effectively control dust.

Water is delivered through a ¼ in. hose to a wet blast head which is attached to the sand blast nozzle. Volume is controlled by a pet cock valve near the water jacket outlet. When the water is turned on, a small quantity of water is sprayed out in a circular pattern directly into the air abrasive mixture as it leaves the nozzle. The force of the compressed air immediately atomizes the water, and, as the dust is produced, it is quickly entrained in the atomized water and precipitated to the ground. When blasting steel, the wet blast is seldom used or needed, but when it is needed, a rust inhibitor such as polyphosphate or chromate is used to prevent flash oxidation of the metal surface.

Dry Blasting

Dry blast cleaning is one of the fastest and most thorough methods of cleaning metal or masonry surfaces. Silica or mineral slag type abrasives generally are used in field blasting operations. Metal shot or metal grit normally is used in process or production line cleaning. Metal abrasives are available in many sizes and are made from iron, steel, and other inorganic materials.

Abrasives for Blast Cleaning

There is a wide variety of specifically processed, high quality abrasives available that meet the requirements of any surface preparation job. The development of these specifically processed or manufactured abrasives, together with advancements in blasting equipment, especially the venturi type nozzle, have been major factors in reducing surface preparation costs. In fact, effective surface preparation today costs less than it did 10 or 20 years ago.

Many factors must be considered when selecting an abrasive. These include: (1) the type of surface to be cleaned, (2) the surface condition prior to cleaning, and (3) the desired degree of surface preparation required, which is usually specified by the user or the adhesive manufacturer.

Literally hundreds of abrasives are available, and each particular one has its own qualities. The qualities include such things as chemical composition and hardness, particle shape and size, availability and price, etc. In selecting an abrasive the following consideration should be reviewed:

1) If sandblasting or abrasive cleaning is recommended by the manufacturer of the adhesive for effective bonding, the manufacturer or the ultimate user should be the best qualified source for this information. Usually the manufacturer has made the laboratory and field tests necessary to determine the exact type of abrasive and characteristics that will provide the best surface preparation, for the adhesive to be used in the area under consideration.
2) The chemical composition of the abrasive is another important consideration when selecting an abrasive. Every abrasive used which is of a

different chemical or metallurgical composition than the surface being blasted will leave traces of itself on the blast-cleaned surfaces. This deposit of foreign material must be removed or compatible with the adhesive to be applied. Compatibility includes both chemical and mechanical factors, such as would affect the wetting strength of the adhesive or the presence of acids or alkali salts etc.

3) The shape of the abrasive predetermines the shape of the anchor pattern that will be produced. The round abrasive will produce a peened effect while an angular material will, of course, make a wedge-type indentation on the surface.
4) The hardness, weight, specific gravity, and shape of the abrasive will determine the depth of the anchor pattern. A vegetable-based abrasive, for example, makes no indentation on steel.
5) The size of the abrasive particles is also a factor in the depth of the anchor pattern. Coarse abrasives will make a deeper penetration than fine abrasives. A uniform size is always desirable so that each grain will not only make the same penetration but also will travel in a uniform flow through the plastic nozzle and hose.
6) The natural color of the abrasive is also important in some instances. A light colored abrasive reflects light, whereas a black or dark abrasive will absorb light. For example, when sandblasting inside a tank where artificial lighting is used, a dark abrasive is advantageous. The dust created from a breakdown of the abrasive can hardly be seen because it will not reflect light.
7) The health hazards are necessarily considered when selecting an abrasive, since the creation of any type of dust is considered dangerous. Clean air feeds to the sandblaster's helmet should be used to protect the health of the operator from abrasive-bearing dusts.
8) Re-using abrasives as in the case of contained systems as rooms or cabinets is seldom ever considered with field type operations using abrasives usually costing less than $25 a ton. The reason, although simple, is not always fully understood. Whenever a metal surface is blasted, it is to remove materials on the surface which will be detrimental to a new coating. The removed material becomes thoroughly mixed with the once-used abrasives. The reuse of the mixture of the abrasive and the removed materials from blasting results in the same foreign materials that were previously removed being redeposited on the surface. The removal of foreign materials plus the rescreening to a usable size of a once-used abrasive would cost more than new abrasives in most field operations.

Types of Blasting Equipment

The most common types of blast cleaning equipment are pressure and suction machines used for wet or dry blasting. A combination vacuum and pressure machine is manufactured which recirculates the abrasive and, at the same time, reduces dust problems. This equipment has limited field application because it is relatively slow compared to pressure machines. Also available are centrifugal wheel machines which use metal shot or grit as abrasives. Wheel machines are used by shop fabricators, pipe plants, and foundries and are not generally found in field operations.

CHEMICAL CLEANING

Degreasing, alkaline cleaning, and etching are methods of chemical cleaning used on a wide variety of adherends. In discussing surface preparation, general and specific steps for the more common adherends are considered.

Degreasing

The first step in cleaning is degreasing or precleaning. Degreasing dissolves and removes organic materials such as oil, grease, and wax. This can be accomplished by: (1) wiping with solvent, (2) dipping or immersing the parts in solvent solutions, or (3) spraying the parts with solvents. The solvents most frequently specified are:

1) MEK (methylethyl ketone), acetone, toluol, and naptha for hand-wipe methods
2) Trichloroethylene or perchloroethylene for vapor degreasing
3) Mixtures of the above for immersing or dipping.

Degreasing not only removes organic materials but also minimizes "carry-over" or overloading of the alkaline cleaner solutions with contaminating organic soils. Degreasing alone is not considered sufficient cleaning for support of metal bond deoxidizing which follows this process.

The most common method of degreasing is to dip a dolly or basket of parts in an emulsified solution for five minutes with the solution at a specified temperature and under agitation.

Alkaline Cleaning

The next step is to immerse the parts in a light- or medium-duty alkaline cleaner (preferably non-silicated) solution. The alkaline cleaner solution should be continuously agitated. The purpose of the alkaline bath is to remove foreign matter such as dirt, metal chips, oil, and other contaminants that often adhere tightly to the metal surfaces and are not removable by degreasing operations.

The alkaline cleaning operation is normally operated at 170° to 200°F and is dumped and recharged when the maintenance additions equal the original charge. After immersion in the alkaline cleaning tank, the parts are rinsed in a continuous overflowing water rinse or spray rinse at ambient temperature until all chemicals are rinsed from the part surface. Care should be taken not to allow chemicals to dry on the surface of the parts.

Etching

Chemical etching involves immersion of the adherend with an active solution which has the power to dissolve a part of the adherend surface, or change it in such a way that the adherend surface becomes chemically active. Details of the etching process appear under the discussions of specific adherend surface preparation.

Surface Preparation for Aluminum

Alkaline cleaning should never be the final treatment on an aluminum surface. The final process is an acid bath commonly referred to as "metal bond etch." The etch removes the residual oxide film from the aluminum surfaces. All aluminum alloys require etching prior to adhesive application to obtain the highest quality bonds. The parts are etched by immersion in a sulfuric chromate solution made of: (1) sodium dichromate, sulfuric acid, and water; or (2) chromic acid, sulfuric acid, and water. A commonly used ratio is as follows: 30 parts by weight of water; 10 parts by weight of sulfuric acid (1.84 specific gravity); and 1 to 4 parts by weight of sodium dichromate ($Na_2Cr_2O_7 \cdot 2H_2O$—technical grade). Solution temperature should be maintained between 145°F and 160°F, and the parts should be immersed for eight to ten minutes under continuous agitation. The temperature for either solution should not exceed 170°F. After etching, two rinses are commonly used. The first is an ambient temperature water spray rinse followed by an ambient or warm immersion rinse. All chemicals must be thoroughly rinsed from the aluminum surface to insure good bonding results. Water used in the etch solution and rinses must be clean and oil-free ("clean" may be defined as deoxidized, demineralized, or untreated, providing it contains less than 50 ppm total solids).

Immediately after rinsing, the etched surfaces should be observed for "water-break" freeness. If the part has been adequately cleaned of oils, greases, and wax, the final film of rinse water should remain in a solid film on the entire part surface without breaking into droplets, pools, or trickles. Such a test does not indicate the presence of materials that absorb water—merely the absence of a greasy surface.

Final operation is to force-dry the parts with warm, filtered, clean air up to a temperature of 160°F maximum (above 160°F can cause detrimental re-oxidation of the part surface).

Conversion Coating. Conversion coating of aluminum prior to bonding for structural strength is normally done using chromic acid anodizing. Most chemical conversion coatings (clear or colored) are not inherently strong enough to achieve good bond strengths. The sulfuric acid anodizing process produces a coating that is normally too brittle to exhibit good peel strengths. Sealed chromic acid anodizing normally will not allow good surface "wetting" by the adhesive, producing "adhesive failure" to the adherend. Unsealed chromic acid anodizing has been used most successfully. After normal degreasing, alkaline cleaning, and metal bond etching, parts are anodized in chromic acid using 38 to 40 volts for 50 minutes. It is important to apply voltage using a programmed voltage increase of approximately 8 v/min; this also applies to the reduction of voltage after the anodizing operation as well.

After anodizing, rinse thoroughly as pointed out earlier, using only clean water at ambient temperature. Do not use warm or hot water because sealing will be initiated. Parts are then dried, using clean filtered air at ambient temperatures up to 140°F maximum. The anodized surface should be protected from unnecessary abrading to avoid "powdering" the coating.

Surface Preparation for Stainless Steel

The following method can be used for surface preparation of most stainless steels:

1) Wipe and rinse with acetone
2) Rinse with distilled water
3) Immerse for 2 min in a solution consisting of 4 oz of trisodium phosphate per gal of water at a temperature of 140° to 160°F
4) Rinse in distilled water

5) Immerse in a chromic acid solution consisting of 4 oz of chromic acid per 100 gal of water at a temperature of 160° to 180°F for 2 min
6) Rinse in hot (140° to 160°F) distilled water
7) Rinse in cold distilled water
8) Dry in a warm oven or by blowing dry clean air on the substrates.

Surface Preparation for Mild Steel

The following cleaning method can be used for preparing the surface of mild steel:

1) The metal surfaces should be completely degreased by solvent wiping, by vapor degreasing, or by various proprietary systems
2) Sandblast or wire brush the steel surfaces
3) If sandblasting or wire brushing is not practical, the steel surfaces should be solvent degreased in hydrochloric acid, 83.3 percent by volume, or phosphoric acid; 12.5 percent by volume for 2 min at 180°F
4) Remove and spray rinse clean with water
5) Dry immediately in an oven at 140°F
6) Coat or apply adhesive to start the bonding processes as soon as possible.

Surface Preparation for Copper Alloys

For copper alloys, including brass, the surface may be prepared as follows:

1) Degrease
2) Dip for 1 to 3 minutes at room temperature into a solution of 42 percent aqueous ferric chloride (15 parts by weight), concentrated nitric acid with a specific gravity of 1.41 (30 parts by weight), and distilled water (197 parts by weight)
3) Deionized water rinse
4) Dry at room temperature with filtered jet air stream (using air above room temperature results in heat oxidation)
5) Bond immediately, or
6) Prime or conversion coat.

Metallic surfaces are more prone to air corrosion than nonmetallics and must be processed immediately after surface preparation. For this reason, primers or oxidation resistant coatings are employed to protect the freshly cleaned surface if the bonding cannot be done immediately. These coatings are much less subject to some types of contamination, and often permit solvent wiping prior to bonding. Minimum handling and protective measures against contamination are emphasized following surface preparation. Other methods of surface preparation are undoubtedly available which will result in similar or improved bondability over the processes herein set forth. It should be noted that one particular type of surface preparation may result in optimum bond strength only when employing a specific adhesive. Surface "roughness" or finish resulting from the method of preparation affects the bond strength to the degree that a specific adhesive wets that surface. Therefore, it must be emphasized that whenever possible the adhesive manufacturer's recommendations for preparing the surface should be followed.

Surface Preparation for Elastomers

The following will show some methods of preparing elastomer type surfaces such as rubbers:

1) Silicone sponge should be thoroughly solvent cleaned and force dried. Silicone stock should be thoroughly solvent cleaned followed by immersion in concentrated sulfuric acid for five seconds at room temperature.
2) Neoprene and polysulfide sponge should be cleaned with a solvent (aromatic) and force dried. Neoprene and polysulfide stock should be solvent cleaned (aromatic), sanded with 320 grit and solvent cleaned again.

Rubber is usually etched by immersing it in an acid solution. This results in what is commonly known as "cyclizing" of rubber which causes minute crack-like fissures on the surface of the rubber.

Different types of rubber require various concentrations and types of acids and different acid etch exposure times for optimum surface preparation. Therefore, each specific type of rubber should be evaluated for the proper surface treatment. Over-etching of rubber by a too reactive etching solution or too long an exposure time may result in fissures of a magnitude that will cause cohesive failure of the rubber at a value lower than the adhesive bond or cohesive strength of the adhesive.

In all cases, mold-release agents that aid removal of the molded product must be thoroughly removed before bonding is attempted.

Surface Preparation for Fluorocarbons

Materials such as fluorocarbons (TFE and FEP), acetals (Celcon* or Delrin†), and polyethylene require special surface preparation procedures before bonding. Not only must surface contamination be removed but the passive surfaces of these materials which normally resist adhesive bonding must be rendered active and bondable. As mentioned earlier, a contaminant such as grease or oil on the adherend surface will render the etching solution ineffective, acting as a barrier through which the solution does not penetrate effectively. Fluorocarbons such as TFE have a surface layer of fluorine atoms which are almost completely resistant to any adhesive known in today's state of the art. The fluorine atoms on the surface of TFE (polytetrafluoroethylene) are removed by a special treatment with a powerful etching solution of metallic sodium in liquid ammonia, leaving a layer of highly bondable carbon atoms exposed. All traces of etching solution must be removed before proceeding with the adhesive bonding. Surfaces not being etched should be masked with tape. An effective etching solution consists of 10 to 20 grams of metallic sodium per liter of liquid ammonia; however, this type of treatment should be performed by experienced and qualified personnel.

Metallic sodium reacts violently with water and is generally stored in kerosene or vacuum packed. Precaution against fire must be constantly maintained. The TFE surface to be etched should be wiped with a solvent such as acetone just prior to etching to preclude any excess moisture on the surface. An excessive amount of moisture will neutralize the etching solution, rendering it ineffective. In the process of etching large surface area TFE parts, this moisture factor becomes significant. High humidity in the work area can also be detrimental. To overcome this problem at least two things can be done:

1) A sufficiently large volume of etching solution in relation to the surface area may be used

*Reg TM, Celanese Corporation of America.
†Reg TM, E. I. du Pont de Nemours and Co.

2) A more concentrated solution may be employed.

The former method is preferred. The TFE parts may be dipped into the etching solution, or the etching solution may be poured into a container holding the TFE part. In either case, care must be taken to prevent splattering of the etching solution. This "splattering" results in an unevenly etched surface. Protective face shields and rubber gloves are necessary to guard against burns. Equally essential is a hood over the tank to carry away ammonia fumes.

The time required for etching is short, between 3 to 5 seconds. Over-etching by using a stronger solution or too long an exposure in a standard solution results in a black, crust-like surface. A color code has been established which specifies a permissible range. After exposure to the etching solution, the excess solution must be removed from the TFE part. If small parts are being dipped, the parts must be removed from the etching solution and dipped into an alcohol such as butyl alcohol and then water rinsed.

For large TFE parts being etched, the etching solution should be drained following etching; an alcohol is then used on the parts to neutralize the etching solution. The alcohol should be drained, and water used for a thorough rinse. The alcohols contain some water but are much less reactive with the etching solution than pure water so that upon contact with the etching solution, reaction is much less violent and therefore safer. This method is more desirable because less ammonia fumes (which burn the eyes) are generated even though an additional step in the process is necessary.

Finally the etched surface should be wiped with clean wiping cloths and water to remove the surface film deposited by the neutralization process. The etched TFE surface should be solvent wiped just prior to bonding. Protective covering is recommended to protect the prepared surfaces against contamination. Protection against sunlight is also recommended. Bonding within 30 days of etching has been recommended.

Other TFE etching solutions are commercially available. Except for etching small TFE parts such as wire or tubing which may be dipped into the solution having a relatively small area exposed to the atmosphere, these solutions have not been found practical. These solutions consist basically of naphthalene, tetrahydrofuran, and metallic sodium, and are very susceptible to moisture contamination. Under controlled humidity conditions such as an inert atmosphere, methods not currently used might be adaptable to etching TFE parts with a large surface area.

Surface Preparation for Acetals

The acetal resin formulated material such as Celcon* and Delrin† presents similar problems to TFE for bonding. The molecules of the surface must be rendered active and bondable. Surface contaminants, particularly grease or oil, must be removed from the surface to be treated. This may be accomplished by solvent wiping. Concentrated acids such as sulfuric or hydrochloric will effectively etch the surface during brief exposure times at room temperature. Additional bond strength may be obtained by the application of a primer to the etched surface. A five minute immersion period in concentrated hydrochloric acid at room temperature effectively etches the acetal surface. Approximately one-third to one-half a mil per minute of acetal surface is removed during the

*Reg TM, Celanese Corporation of America.

†Reg TM, E. I. du Pont de Nemours and Co.

etching process. Surfaces not requiring etching should be masked with vinyl tape. After etching, air drying parts for four hours at room temperature has been found to result in optimum bond strength with most acetal adhesives. Numerous other methods such as forced drying for various time periods, solvent wiping, and a combination of both (following etching) did not produce comparable data to the four hour air drying method.

When acid etching acetals, it is recommended that the acid be circulated using a glass rod to move fresh acid into contact with the acetal. The acetal acts as a base and tends to neutralize the acid. This acid etching process might be called the acid test. If sufficient machine or molding stresses are present in the acetal, cracks will readily become noticeable following acid etching, and parts will be useless. The solution to surface cracking lies in stress relieving the acetal prior to etching. Although acid etching of the acetal produces a good bondable surface for numerous adhesives, even greater bond strength may be obtained using a specifically developed primer on the etched surface, but the primer must be applied in a very thin coating for optimum results.

Heat formed or machined acetal parts should be stress relieved not only to prevent cracking from acid etching but also to guard against dimensional change during the primer cure (if used) or elevated temperature cure of the adhesive during bonding operations. Fixturing is recommended in such cases where sufficient heat is employed to affect this thermoplastic material's dimensional stability.

Surface Preparation for Polyethylene

Some success has been obtained in adhesive bonding of polyethylene if it is solvent cleaned or acid etched to remove low molecular weight fractions from the surface. In addition, applying a flame to the surface will oxidize the surface and permit bonding.

MISCELLANEOUS CLEANING METHODS

Soaps and Detergents

Soaps and detergents reduce the surface tension of water below that of the oil layer, emulsifying the oils. Caustic soda forms solutions with the oil layer. Soaps, detergents, and caustic soda may be applied by scrubbing, dipping (ultrasonic agitation), and spraying. The advantages of these materials are that they are cheap, nontoxic, and easy to handle; one disadvantage of these materials involves their reactions with some metals, forming additional contaminants. Consequently, they should be used hot.

Hand Tools

Hand tool cleaning is accomplished by tools such as scrapers, wire brushes, and chipping hammers to remove old paint, rust, scale, and weld splatter. This slow method should be used only when the area is small and the coating or the adhesive to be used is compatible with this type of surface preparation.

Pneumatic and Electric Tools

Power cleaning tools are used where speed and low costs are desired in preparing the surface. These include air or electrically driven wire brushes, abrasive discs, chipping hammers, and scalers.

These tools are not recommended if the surface to be cleaned is pitted or if a high performance bonding or protective coating is to be applied. Neither power tools nor hand tools will remove rust from pits and neither will produce the surface profile required to anchor most of the new coatings without excessive metal removal.

Flame Cleaning

Heat and flame produced by burning a petroleum product are sometimes used to remove contaminates and some other foreign materials from metallic surfaces before bonding. A preheating machine will burn off deposits of oil, grease, and old paint.

Pickling

During production of ferrous metals such as steel, oxide films are formed on the metal surface. These films must be removed prior to coating. This can be done in galvanizing plants by pickling, a process by which the metal is immersed in aqueous solutions of sulfuric, hydrochloric, phosphoric, nitric, or hydrofluoric acid, depending on the nature of the oxide and of the metal to be cleaned. Removal of these oxide films by pickling offers a sufficient degree of surface preparation for some adhesive bonding jobs.

Hydraulic Cleaning

Water blasting is one of the oldest methods of removing dirt, rust, scale, and other contaminants. High pressure pumps and a water supply are basic to this cleaning method. It is used in marine and industrial applications where environments preclude the use of abrasives. Water blasting eliminates dust and explosive hazards but will not remove such materials as pit rust or tightly adhering mill scale, and it does not produce any anchor pattern. When abrasives are also introduced into this cleaning method, the process is usually referred to as hydroblasting, and some special units can operate at pressures up to 5000 psi.

Miscellaneous Cleaning

There are many other types of cleaning including tumbling and vibratory cleaning. These special devices have a few field-type applications.

Preparation of Wood as an Adherend

In the preparation of wood as an adherend the following steps should be taken:

1) The wood should be dried to its normal moisture content consistent with the environmental use of the assembly (generally from 5 to 8 percent)
2) Remove surface contamination by plane, sanding, or machining
3) Clean debris from surface with a brush, vacuum, or air, or wipe it with a solvent-moistened cloth
4) Determine the need for filler, sealer, or primer to offset excessive porosity
5) Minimum handling using protective measures.

Since wood is hydrophilic and porous, it swells with an increase in the relative humidity and shrinks with a decrease in moisture. Besides the basic ingredients of cellulose and lignin, wood contains resin, wax, and oxidation products, varying with different types of wood. Softwoods may contain high resin content which oxidizes on the surface with time, making it harder to wet the surface with adhesive. Under these conditions, best results are obtained by bonding freshly prepared joint surfaces with little or no time lapse between the preparation of the surfaces and subsequent application of the adhesive.

Overheating wood results in a poor bonding surface. The surface becomes hydrophobic.

Surface roughness or smoothness requires the consideration of adhesive wettability. This can be corrected with extenders, diluents, or some other modification of the basic film forming substance or resin.

Solvents for Cleaning Acrylic Plastics

Methylmethacrylate or acrylic, commonly known as Plexiglas* or Lucite,† are most often bonded using solvent-type adhesives which attack the molecules on the surface.

Solvents such as naphtha, Freon,‡ or carbon tetrachloride are used to clean the surface prior to bonding. Solvents such as acetone or methylethyl ketone will actually bond two pieces of acrylic by attacking the surfaces and in so doing fuse them as one unit. Numerous acrylic cements are commercially available. The low viscosities of the cements require accurately mated bonding surfaces. Cement viscosities may be increased by dissolving acrylic shavings in the solution which not only results in improved handling characteristics but also reduces shrinkage (stress) and permits fillet-type application.

---

*Reg TM, Rohm and Haas Company.
†Reg TM, E. I. du Pont de Nemours and Co.
‡Reg TM, General Motors Corporation.

REFERENCES

1. *Steel Structures Painting Manual* (Pittsburgh: Steel Structures Painting Council, 1955).
2. *Ibid.*
3. *Ibid.*

chapter 5

methods of bonding

In an assembly of components via the adhesives method, the availability of so many different methods of adhesive bonding can be used to advantage. But this multiplicity can also be a problem to a manufacturer. His choice of method of application can, for example, restrict the degree of freedom to be exercised in designing the end product. Also, the method of application can affect the selection of materials that can be used in manufacturing the product, the quality or performance of the product, and, very often, assembly costs.

With so much at stake, the manufacturer unavoidably must spend considerable time becoming knowledgeable of materials and methods. Elemental to the ultimate choice of method are the following:

1) The size and shape of the parts to be bonded
2) The specific areas to which the adhesive is to be applied
3) Number of assemblies to be produced
4) The required production speed
5) The viscosity or other characteristics of the adhesive.

MANUAL ADHESIVE APPLICATION METHODS

The so-called manual methods of adhesive application combine the advantages of simplicity and low capital investment in equipment. Several manual methods are so basic and simple that they may not receive the attention they merit. It is possible that many assembly lines are being set up with more sophisticated equipment than the operation really requires, simply because manual methods were prematurely and undeservedly judged to be inefficient. Manual applying methods that have proven themselves in many operations, even on high-speed production lines in extremely busy plants, are brushes, flow brushes, oil cans, squeeze bottles, spatula, silk screen, hand dipping, manual glue guns, and various types of rollers.

Brush

Probably the most widely used of all methods, brush application is simple. It requires minimum clean-up and waste. It permits application of adhesive to only a small segment of a surface, when such is desirable and masking is impractical. For small or irregular parts or low volume production, brush coating is particularly effective. One problem with brush coating is that uniform film thickness is somewhat difficult to achieve, and the operator's care and judgment will determine how well the adhesive is applied. However, many operators acquire an amazing degree of proficiency over relatively short periods of time.

Flow Brushes

Hollow brushes through which adhesive is fed—with or without air pressure—offer a higher rate of production, produce a more uniform coating, and cut down waste. In its simplest form, a flow brush carries adhesive behind the bristles or in a chamber feeding down the side of the brush. A more sophisticated type, capable of applying heavier adhesives or sealants, is fed by air pressure and can provide continuous or intermittent flow. A fingertip control valve on the brush barrel regulates adhesive flow. Some models have three outlets on one supply tank for the simultaneous use of up to three brushes.

Brushes for applying adhesives should be of high quality, made with either all animal bristles or nylon. Ordinary paint brushes tend to fall apart from immersion in the solvents present in many adhesives. Also note that adhesive should be brushed on in a single, smooth sweep, and that a second sweep over the same area is not practical with most types of adhesives. Cleaning of brushes is best accomplished by using a solvent known to be suitable for the particular type of adhesive employed.

Oil Can and Squeeze Bottle

The hand-pump oil can and the polyethylene squeeze bottle are often used for spot application of adhesive. Where it is necessary to apply adhesive inside a blind hole or some other point with limited access, the oil can with a stem of appropriate length can be a simple solution. Tips at the end of the stem can be devised to deliver multiple spots of adhesive. The squeeze bottle can sometimes be used for the same purpose, and also to apply a continuous bead. When using polyethylene squeeze bottles, care should be taken that the adhesive will not affect the polyethylene itself.

Spatula

A simple applying tool particularly effective for hard-to-spread adhesives is the spatula. Knives and trowels, both of which may have notches cut into their applying edges, are variations of the spatula. They are wielded in the same manner. Depth of notches and the spacing between them help regulate the amount of adhesive applied. Holding the blade firmly and at a right angle to the surface is essential for proper adhesive application.

The preferred shape for spatula notches is square; triangular notches present sharp teeth which wear out much faster than the wide contact surfaces of the square notches. Shallow, rounded and closely spaced notches are often used with adhesives of higher liquid contents because notches of such a design permit the adhesive to flow together and thus form a continuous, unbroken film.

Silk Screen Process

If the adhesive is to be applied only to specific parts and not the entire surface of a flat sheet, a silk screen process is often used. Low viscosity adhesives work well because they pass readily through the cloth onto the work. However, only relatively thin films can be applied by this method. By filling the pores in a cloth over areas to remain uncoated and leaving other portions of the cloth unfilled, effective masking is set up. Then the adhesive is poured on top of the cloth and a rubber squeegee used to facilitate flow of adhesive down through the cloth. The squeegee, by itself, is another manual method of applying adhesive. As a general rule, fast drying or tacky adhesives cannot be satisfactorily silk screened.

Hand Dipping

The dipping process is capable of speeding up the coating of relatively large surfaces. Automatic versions are the most satisfactory, but hand dipping can be used to speed up the application of adhesive. The devices used to facilitate immersion of parts into the adhesive bath or trough range from baskets to screens to perforated drums, but in every case coating is quickly accomplished.

Manual Glue Guns

Adhesive application with a glue gun permits the placement and spreading of the adhesive by hand with precision and speed. Such guns are widely used in doweling, mortising, and sealing operations, as well as in assembly work. This discussion, however, is restricted to their use for adhesive application.

The adhesive is fed by hose under air pressure to the gun. By adjusting the air pressure, the speed of adhesive flow can be matched exactly with the speed of application; selection of the proper applicator facilitates application of the adhesive onto the particular surface configuration involved—flat, corner, groove, hole, etc. An almost countless variety of tips is available. The guns can be used to bond on all types of surfaces: cardboard, paper, foil, plastic, metal, wood, felt, rubber—using of course the proper adhesive. Guns are available to dispense adhesives over a wide range of viscosities.

The glue gun (also referred to as a flow gun) may be activated by a trigger device by flexing the gun body, or by pressure on the gun tip, depending on the model. In every case, however, an internal valve opens which permits the extrusion of the adhesive.

Rollers

One type of roller consists of a plastic or hard rubber cylinder attached to some type of handle, and is used primarily to spread heavier or tackier adhesives which have been previously deposited on the surface to be bonded by some auxiliary means, such as spatula or knife. It can be used on all types of surfaces, but the only control over thickness of adhesive spread is the skill of the operator. Available in various diameters and lengths, the simple roller is far from a sophisticated means of applying adhesives, but simplicity of operation makes it ideal for many applications.

Another type is the ordinary paint roller very often used to apply adhesives onto relatively large surfaces, particularly if conditions call for heavier films. It offers the advantage of serving as its own means of transferring adhesive from container to surface. Adhesives often are applied in this manner for the bonding of plastic laminates to particle board in the construction of kitchen counters, bathroom vanities and similar fixtures.

Paint rollers consist basically of a core and a covering, and their longevity and efficiency depend upon the nature of the materials from which the two basic components are made. A core constructed of fiberboard, while completely satisfactory for ordinary adhesive applications, cannot be expected to stand up against constant immersion in adhesive and solvents as effectively as a core made of phenolic or other plastic material. The covering, which serves as the actual applying surface, may be made from lamb's wool, synthetic fibers of varying length, or foam rubber. Selection of a roller with good resistance to chemicals is imperative when applying solvent-based adhesives.

Pressure-fed paint rollers can be used to apply adhesives of relatively low viscosity. Their chief advantage over ordinary paint rollers is that they eliminate the need to dip the roller into the adhesive to replenish supply; the adhesive, instead, is fed from inside the roller. Tanks of varying capacities are available, and rollers come in a wide range of sizes.

Still another type is the flow roller. As in glue guns, adhesive is supplied to the flow roller by a passage from the rear. Because the adhesive supply is totally enclosed, the flow roller is often specified for applying solvent type adhesives whose effectiveness can be severely impaired by any loss of solvent to the air. The roller can be of nearly any construction and often is knurled to transfer more adhesive per rotation. Edged veneering and wood strip gluing are examples of usage.

MACHINE APPLICATION METHODS

The application of adhesives with machines designed for the purpose generally speeds up or simplifies assembly. Selection of the particular machine method best suited to a given operation is not particularly complex despite the number of distinct methods. On the other hand, the selection of the best variation of the basic machine method can be difficult. First-hand familiarity with what is available can be extremely valuable to a manufacturer.

Bench Roll Coaters (see Fig. 5–1)

Savings of both applying time (20 to 35 percent is common) and of adhesive waste (from 20 to 35 percent—much more in some cases) have been reported

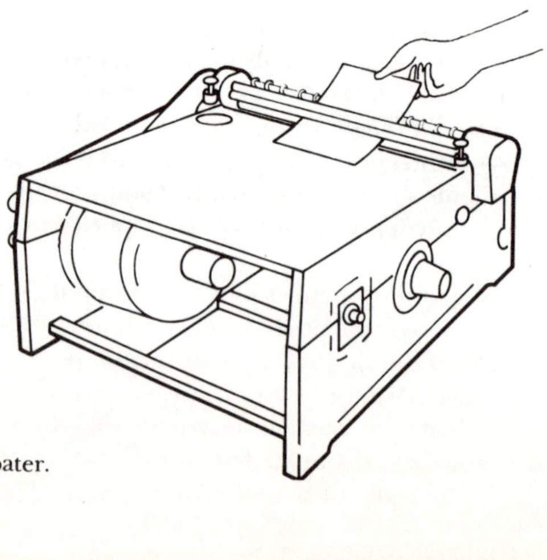

Fig. 5–1. Bench coater.

by firms using bench coaters in place of hand brushing. As indicated earlier, hand brushing remains a practical method of application for low volume production. But when production reaches a rate of 200 to 300 pieces per day (depending on the size of individual pieces), a changeover to bench coaters should be considered. Top production rates ranging between 5,000 and 12,000 pieces per day are possible when bench coaters are used, particularly if good work flow conditions exist and suitable racks or conveyors are employed.

There are many different models of bench coaters available, making it possible for the user to meet his own needs. The most popular size of coaters, measured in terms of applying roller width, are 4, 7, 8, 10, 12, 14, 15, 20, and 26 inches. In general, most models include a motor which drives both adhesive and feed rollers, a pump that moves the adhesive from the tank's lowest level to the applying roll, and various arrangements of transfer rolls and doctor blades for control of adhesive thickness. In almost all cases, the adhesive transfer system and the tank are completely enclosed to keep solvent evaporation and loss at a minimum. This feature, combined with the unavoidable waste of adhesive by even the most careful brush operator, assures substantial adhesive savings. Some equipment, not fully enclosed, is equipped with tanks that dip fresh solvent into the adhesive reservoir to replenish solvent lost by evaporation. Whenever solvent-based adhesives are used, explosion proof motors and switches should be specified.

Some bench coater models employ feed rollers of the so-called open-end style. This allows work as large as can be handled or guided to be run through the machine. Furthermore, it permits variations in the patterns of adhesive application so that adhesive will be applied only on the part that passes through the applying rollers. Models without this open-end feature limit the width of the work that can be run.

Bench coaters having removable feed pressure rolls make it possible to apply adhesive to the ends or edges of large objects. With clear access to the applying roller, slabs of foam material, for example, can be held in a vertical position and passed over the applying roller. The objective is to simplify butt joining of such objects.

Most manufacturers of bench coaters provide or will supply applying rollers with smooth surfaces for thin applications and increasingly coarse grades of knurled rollers for heavier coatings. Literally dozens of variations of groovings including coarse to fine horizontal and spiral patterns can be supplied to meet any special condition or adhesive viscosity.

Floor Mounted Roll Coaters (see Fig. 5–2)

More adhesives and other coatings are applied with roll coaters than by any other single applying method. One reason is that roll coaters are extremely efficient for use with large-sized webs and individual flat sheets or panels.

Waste of material is very low. Some manufacturers claim that 98 percent of the coating material put into the machine is effectively coated onto the work in process. Because the value of the machines are realized most with continuous and volume production, floor-mounted roll coaters are used in many of the larger plants.

Variations of type are numerous: kiss roll, pressure roll, reverse roll, and specials. Application is common to all by contact of the web, sheet, or panel

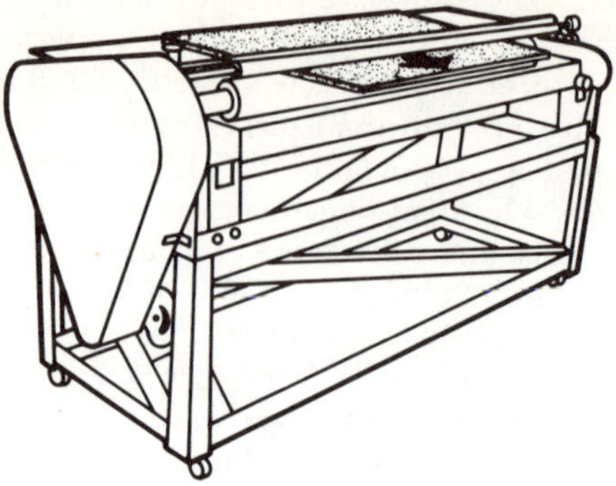

Fig. 5–2. Floor mounted roll coaters.

with a roll surface. Once selection of the proper type is made, the user must choose a particular make and model. Price will be a factor, but so should the accuracy with which the coater, and particularly its controls, are built. The precision-made unit will keep problems of coating uniformity at a minimum, and also provide maximum adhesive mileage.

Materials processed through roll coaters in continuous webs fall into four general classifications: (1) non-woven webs, (2) woven fabrics, (3) films and foils, and (4) metals.

Materials in web form must have sufficient flexibility to be processed from and to rolls at high speeds. Materials processed in sheet or panel form may be flat or nearly flat in shape and are usually semi-flexible or rigid. Many roll coaters for panels are adjustable for thicknesses up to three inches. Wallboards, door panels, table top cores are typical panel types. Roll coaters are used to apply adhesives or coatings to many materials, e.g., paper, paperboard products, plastics, synthetic rubbers, cloth, wood, composition materials, and metals.

A variety of formulations can be applied by roll coaters. Some commonly used are: solvent resin solutions, caseins, and other cold glues, as well as latex, lacquers, primers, paints, varnishes, and waxes.

Coated webs eventually are processed into numerous products, but typical are: pressure-sensitive tapes, decorative wrapping papers, flexible packaging materials, vinyl, rubber, and linoleum floor coverings and such metal products as venetian blinds, aluminum clapboard sidings, awnings, and toys. Frequently, sheets of leather, canvas, wood, glass, plastic, and metal are coated with adhesive prior to use in a wide variety of industries.

No single model or type of roll coater can provide efficient service for all of the variables of material surface, coating formulations, and end use. To accommodate the various combinations that emerge out of these primary factors, each of three major types of roll coaters (kiss roll, pressure roll, and reverse roll) has its uses and characteristics.

The "Kiss" Type Roll Coater (see Fig. 5–3). The "kiss" type roll coater, so named because of its light contact with the applicator roll, is more common to web applications. Since no back-up roller is used, the contact is controlled by web tension. It frequently creates rib marks or a rippled pattern on the coated surface. The transfer of coating to the web is accomplished by surface

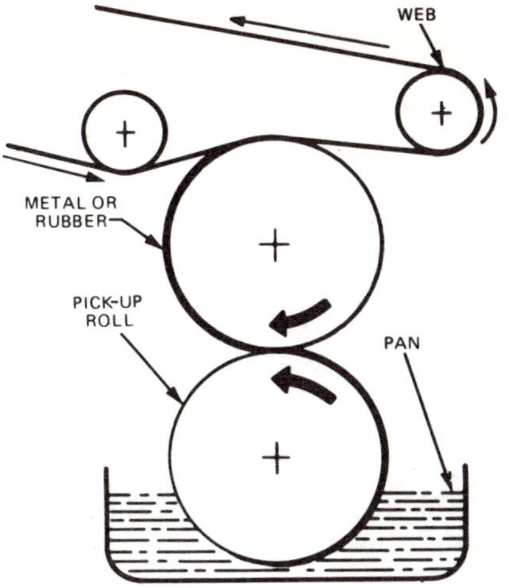

Fig. 5–3. Pan fed "kiss" roll coater.

contact with a roll traveling either in or opposite to the direction of the web. When the roll runs with the web direction, a split of the coating film results, but the web receives a fairly uniform coating. Usually a premetering device such as a doctor blade is used for control. Depending on the ability of the coating material to level itself out in use, subsequent smoothing operations may or may not be required. Kiss coaters are more often used as adhesive coaters than as appliers of finish materials. It is at its best when used for coatings with good flow characteristics and relatively slow drying rates, and for adhesives that will be sandwiched between two layers.

Pressure or Squeeze Roll Coaters (see Fig. 5–4). Pressure or squeeze roll coaters, sometimes called nip coaters, also leave rib marks, but with the use of resilient covered rolls, can contribute greater coating penetration into open

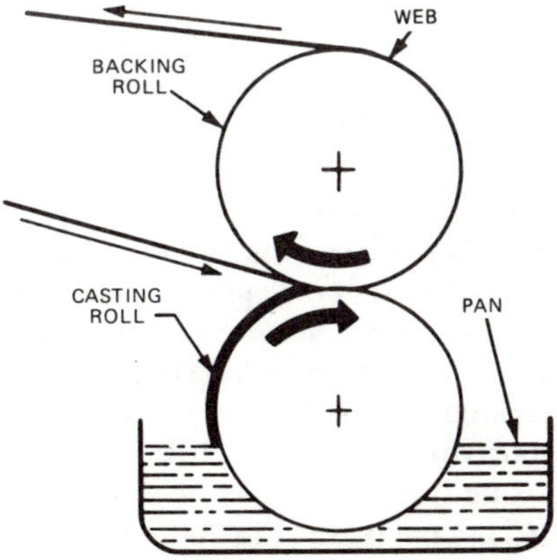

Fig. 5–4. Pressure or squeeze coaters.

web structures due to the greater area of nip contact and longer nip dwell. Both the web and the coating materials pass through the rollers, and the excess coating is squeezed back. This type of coater is widely used for application to both webs and rigid panels, but is particularly suited to panels, as the direction of the rolls helps move the panels through the machine. If desired, the coatings may be applied both on the face and on the back of the material simultaneously. These are termed double coaters. Pressure coaters are often used for fairly thin solutions where nip patterning is less apt to occur, and with all viscosities where appearance is not a critical factor, as in adhesive application. On the debit side, because both the coating and material pass through the metering nip, pressure roll coaters tend to vary the thickness that they apply with variations in the thickness of the passing material.

Reverse Roll Coater (see Fig. 5–5). Of all roll coaters, the most popular for web precision coating is the reverse roll coater. The transfer roll, running in reverse to the surface being coated, provides more control of coating thickness. It premeters coating, thus ensuring uniformity of coating independent

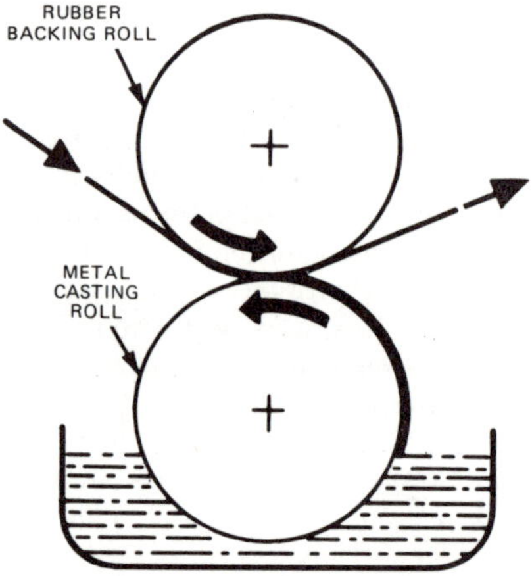

Fig. 5–5. Reverse roll coater.

of surface irregularities or variations of thickness across the web. The number of roll arrangements possible is one reason why so many reverse roll coaters are in use. The coating thickness is controlled by varying the speed of the surfaces to be coated, and by operating the coating roll in a fixed, but adjustable, spacing from a metering roll which, in turn, is cleaned by a doctor blade. When this roll is rotated in the reverse direction from the coating roll, an accurately metered, smooth coating results. In many cases, a backing roll is used to apply sufficient pressure against the coating roll to ensure transfer of the coating by wiping. This greatly reduces the ripple pattern. However, if the coating material is not sufficiently free flowing, rib marks may be visible.

Two-Roll Reverse Roll Coater (see Fig. 5–6). Reverse roll coaters are similar to the squeeze roll type, but some additional control is offered by having the transfer roll running in reverse. Adding a third roll and doctor blades for

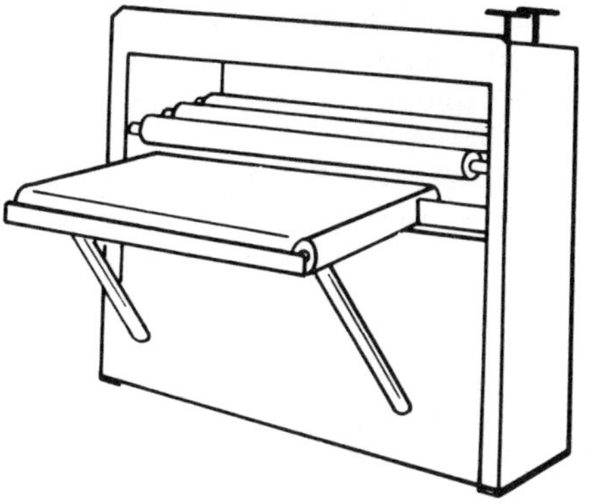

Fig. 5–6. Two-roll reverse roll coater.

metering and cleaning give much greater control. The coating in this type is fed into the nip of the transfer and metering roll using edge dams to retain the liquid material. Four-roll types have the advantage of better visibility with web materials. The fourth roll is a feed roll running in a pan, and this extra roll partially premeters and reduces the possibility of air entrapment in the material, and assures high-quality coating.

In reverse roll coaters, applied wet film thicknesses can range from 1 to 20 mils. The three- and four-roll coaters can handle thicker materials with viscosities ranging to 200,000 cps.

Dip Roll Coater (see Fig. 5–7). Probably the simplest of all basic types, the dip roll coater is used extensively to coat both sides of a web simultaneously. In most cases, the coating is one of low viscosity. Impregnation or saturation coating requirements are often best accomplished by dip-coating equipment. In operation, the web travels vertically down and around a roll that is either partially or completely immersed in the coating formulation. This option

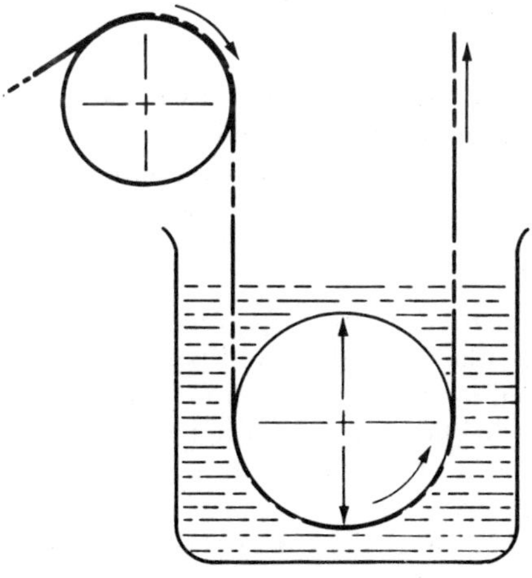

Fig. 5–7. Dip roll coater.

permits control of application to either one or both faces of the web. Many modifications can be made. Scrapers, metering rolls, and doctor bars—in each case to help control the amount of coating applied—are some of the most common and useful modifications.

The Engraved Roll Coater (see Fig. 5-8). A very accurate control of coating uniformity is provided by the engraved roll coater. The amount of adhesive or coating applied is measured by the style and depth of engraving or indentations in the coating roll. This type is most advantageous for extremely low coating weights and may be used for aqueous or solvent coatings. Various modifications, including doctor bars and offset rolls, may be used.

The Air Blade Coater (see Fig. 5-9). Primarily associated with web processing, the air blade coater is basically a form of the kiss coater in the sense that the solution is applied on a material as the material touches the coating roll. But added to the unit is a backing roll around which the web passes. An

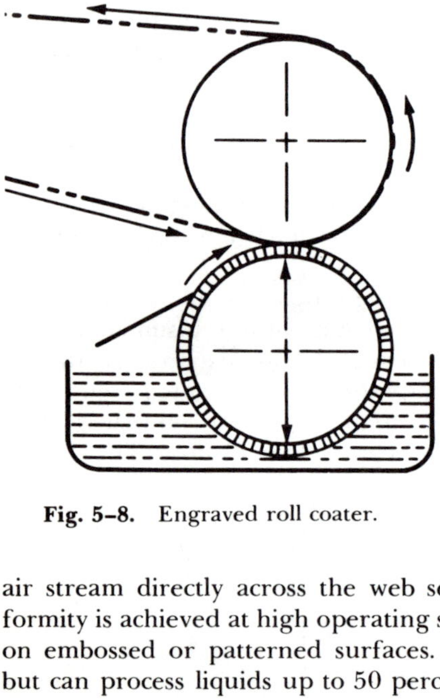

Fig. 5-8. Engraved roll coater.

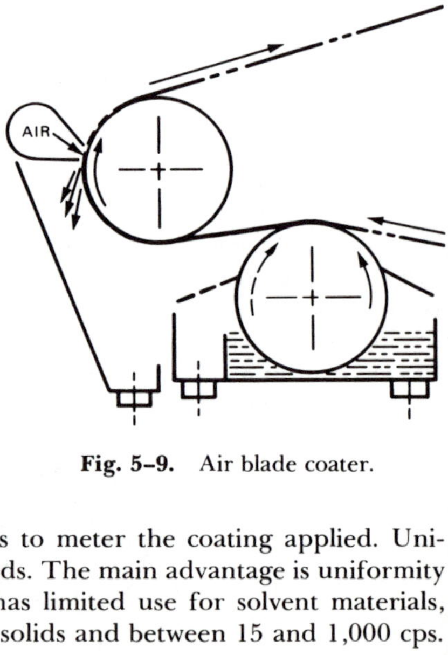

Fig. 5-9. Air blade coater.

air stream directly across the web serves to meter the coating applied. Uniformity is achieved at high operating speeds. The main advantage is uniformity on embossed or patterned surfaces. It has limited use for solvent materials, but can process liquids up to 50 percent solids and between 15 and 1,000 cps.

Flow or Curtain Coaters

A relatively new method of applying adhesives, coatings, or finishing materials to wood, metal, plastic, leather, and paper surfaces is variously referred to as flow coating, veiling, or curtain coating. Basically, a flow coater consists of a head with a slot in the bottom through which the coating or adhesive is poured onto the surfaces of the objects to be coated as they pass beneath the head by means of a conveyor. A pump serves to deliver the coating material from a reservoir to the coating head. Across the width in the bottom of the head is an adjustable orifice to regulate the thickness of flow of the curtain-like film extending over the entire width of the machine. All of the coating material that does not make contact with the surface of the passing objects flows into a return trough and back into the reservoir, accounting for a very important advantage of flow coaters over competing methods – the conservation of coating material. Also a major advantage is the coating uniformity which can be achieved even when surface irregularities are present on the objects being coated. The flow coater will apply the same thickness of coating on the hills as well as the valleys of the objects.

It is essential to understand that in addition to the selection of the orifice opening, the coating film thickness laid down also is controlled by the speed of the conveyor system which passes the objects under the coating head. In brief, the coating thickness applied varies inversely with the conveyor speed. The viscosity of the coating material also is an obvious factor. Therefore, it is possible to deposit exactly the desired thickness from very thin (fraction of a mil) to very heavy (100 mils) by selecting the proper viscosity for the coating material, by adjusting the orifice, and by selecting the conveyor speed. The height of the coating head above the material to be coated is another variable, and although, by itself, it does not affect the final coating thickness applied, it must be selected with care. The head may be raised too high, causing the curtain to become too thin and to "break." The head sometimes is raised to increase the velocity of the curtain at the point where it contacts the passing objects and thereby ensures consistent coating of irregular contours.

One pressure curtain coater has a pressure head which produces a pressure curtain of the coating material. The material is pumped to the head against an entrapped air cushion. Another model with an open head design does not require adjustable orifices to make a curtain. The coating material is pumped into the head weldment and permitted to flow via gravity over a weir or dam. Accordingly, the thickness of material deposited on the workpiece is regulated by the flow rate and the speed of travel. Regulation of the pump and conveyor speeds are the only adjustments necessary to operate the coaters.

SPRAY APPLICATION EQUIPMENT

Materials applied by spray equipment include paint, enamel, synthetic finishes, lacquers, varnishes, waxes, and adhesives. The surfaces to which these materials are spray-applied include wood, metal, plastic, ceramic, glass, leather, fabric, and paper. When other variables are added, such as the viscosity and drying characteristics of the particular formulation being sprayed and the nature of the product being coated, selection of the best suited equipment can provide significant advantages.

The selection of spray equipment over other methods of application is usually predicated on economic considerations. This is particularly true when the objects to be coated are large or of a complex shape. Spraying is fast and provides a means of reaching inaccessible areas easily. Furthermore, the drying time of the coating usually is reduced when it is sprayed. When automated spray equipment is employed, particularly for long runs of identical objects, the optimum advantages of spray application are realized.

The spray application of any formulation, whether its function is bonding or simply coating of a surface, calls for the atomization of a free stream of liquid. This is achieved only when the liquid is made to travel in excess of a certain speed related to the viscosity of the liquid. The higher the viscosity, the higher the velocity must be with which the atomizing device projects the liquid.

Spray equipment can be classified by method: "air spray" and "airless," depending on the technique used to project the liquid into a fine stream (see Fig. 5–10). The conventional is the air spray method in which the liquid is atomized by striking it with a high-speed jet of compressed air. The airless

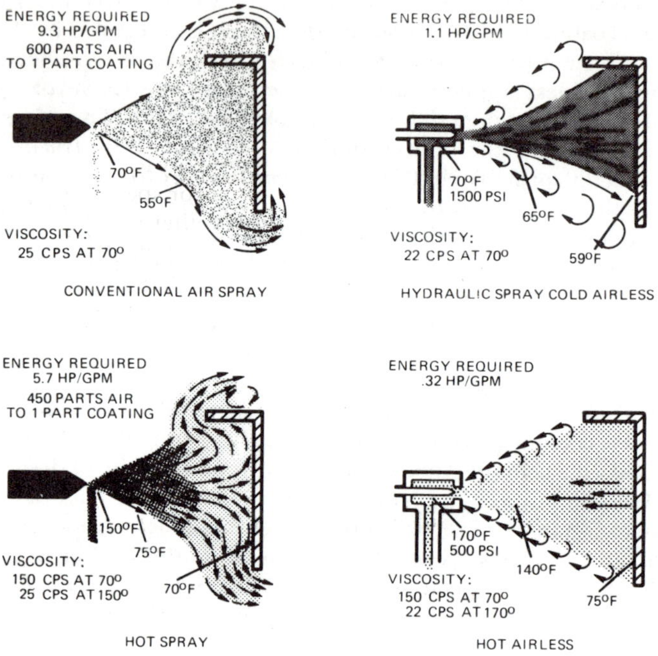

Fig. 5–10. Comparison of four spray methods.

method depends on hydraulic pressure for the atomization of the liquid. The ordinary garden hose is a simple example. Once the liquid has left the nozzle, momentum is the only force that carries it forward because the only material emerging from the airless spray gun is the atomized coating or the adhesive itself. The air between the nozzle and the target resists the motion of the atomized particles, slowing them down appreciably. This serves to reduce overspray. In either case, air spray or airless, the characteristics of the discharging stream can be varied by varying the orifice while holding the pressure constant or by changing the pressure while holding the orifice to one size.

The spray material can be heated prior to atomization to achieve certain desirable advantages. If the liquid is heated above ambient temperature before it is subjected to air spray, the spray application equipment is commonly referred to simply as "hot spray." When hydraulic pressure is used to atomize the heated liquid, the spray equipment is customarily referred to as "hot airless." In either case, elevation of the temperature of the coating material usually lowers its viscosity and consequently decreases the atomization energy required.

This, in turn, makes it possible to apply heavier film and reduce overspray losses. In addition, the use of heat at a properly selected temperature can also eliminate the effects of ambient temperature and humidity, both of which can contaminate the coating by creating atmospheric water vapor condensation on the surfaces.

Air spray, airless spray; heated, nonheated; from these variables emerge four basic spray methods. A comparison of the four, showing patterns of spray and energy requirements, is provided in Fig. 5–10.

Regardless of the type of spray method, at least two basic components are required: a container for the material to be sprayed, and a device for actual application of the spray to a surface. Other essential components for spray systems are pumps and hoses, but the selection of these is dependent on the type of spray method. When materials are to be raised above ambient temperature, heaters and filters are also essential.

The container, usually glass or metal, serves as a supply reservoir and may be of the cup, tank, or gravity bucket type. The last is a container placed at a level above the spray gun to allow the material to flow down to the gun. Cup containers are widely used where comparatively small quantities of materials are to be sprayed. There are two types, classified by the nature of the force which provides flow of the material – suction and pressure.

The device for actual spraying and application of the coating is a spray gun.

Spray Guns

Spray guns are made up of two basic elements: (1) the body of the gun, including valves and actuating trigger, and (2) the nozzle combination (fluid and air).

If the coating material is atomized by air spray, the gun has separate passages for air and material so that the two are mixed and ejected in a controlled pattern. Spray guns may or may not have the container attached. They may be of the bleeder or nonbleeder type; may provide external or internal mix and pressure; may be designed for gravity, suction, or pressure feed; or may be of a special type for so-called automatic operation, principally because objects to be coated are carried past it on some type of conveyance.

In many systems, pumps move the spray material from container to gun at a desired operating pressure. Some pumps are capable of moving as much as 900 gal/hr. Popular with many are special air-operated pumps which move materials directly from original containers such as 55 gal drums. For some airless spray units, pumps with input-output ratios from 20 to 1 to as high as 40 to 1 are available. These are both air and electrically driven pumps, and both gear and piston types.

Two types of hose, air and fluid, are used in spray equipment. Covers usually are made of rubber, although for small, low pressure units, the air hose may have a braided cover. The inner tube of the hose is constructed of special material, formulated to resist the actions of all solvents commonly used in coatings. Ordinary rubber composition liners will not do. Size of hose (expressed by inside diameters) is a significant factor to be considered in selecting equipment. The size governs the air pressure drop at the spray gun.

Manufacturers recommend that end users choose first the basic gun model appropriate to their application, and then the nozzle combination to be affixed to it. In choosing a gun model, the user must consider: (1) production volume,

(2) the kind of handling the gun will have to withstand in this application, (3) the material the gun will spray—not only because the material will affect the working parts of the gun but also because some gun models are specifically designed for spraying certain materials, (4) gun capacity required, and (5) adjustments and controls desired.

In choosing a nozzle combination for the spray gun, the following factors should be considered: (1) nature of material to be sprayed, (2) method of feeding the material to the gun, (3) volume of compressed air available for operation, and (4) the desired texture and decorative pattern of the sprayed material.

The nozzle combination is made up of the fluid tip and the air cap. The choice of a fluid tip nozzle is determined by the speed of the application required and the viscosity of the fluid: the greater the speed of application, the larger the fluid opening required. The choice of an air cap nozzle is determined by the type of materials to be sprayed, the nature of the spray, and the volume of air available. The function of the nozzle is to atomize the fluid. If this is accomplished by the action of air jets which are outside the nozzle combination, it is called an external mix nozzle; if the fluid is atomized inside the air nozzle, it is called an internal mix nozzle. The former is more generally used for production work, for fine finishes, and can be used with all types of fluids. The spray pattern is adjustable. The internal mix nozzle is not fully adjustable; the spray pattern is determined and fixed by the shape of the nozzle. However, the size of the pattern and the degree of atomization is adjustable by regulating the pressure ratio and flow rates of air and fluid to the nozzle. The internal mix nozzle does not require as much air and produces slightly less overspray. It is used primarily for maintenance spraying and for slow drying materials where extremely fine atomization is not required.

Cubic Feet Per Minute (cfm) Requirement. Selection of a nozzle combination (tip and cap) depends on the volume of available air to supply the gun as well as other equipment using air in the plant. Identifying specifications on the nozzles, as furnished by manufacturers, usually denote the cubic feet per minute (cfm) of air used at the required pressure for the type of application involved. If the spraying pressure is increased, the volume of air required by the nozzle will increase, while a decrease in the spraying pressure will decrease air consumption. Finally, the volume of air used by the nozzle generally indicates the speed of application as well as the degree of atomization. The greater the air consumption, the faster the coating material may be applied, and the finer a given amount of material can be atomized.

Spray Patterns and Nozzles. As noted earlier, an external mix air nozzle is adjustable for applying spray patterns varying from round to fan-shaped, including all intermediate shapes. For any special pattern not falling under this grouping, a special nozzle will be needed. This service is available from manufacturers. If airless spray units are employed, it is necessary to change nozzles in order to change the pattern of spray. Adjustable nozzles for airless units are not available at the present time.

Special Purpose Spray Guns. Multiple component materials such as catalyst-type resins and two-part adhesives are sprayed by a two-component spray gun or a plural component gun. A basic difference between the two is that the two-component gun mixes the components within the spray gun, while the plural component gun mixes outside the nozzle.

METHODS OF COMPLETING ASSEMBLY

Once an adhesive has been properly applied to one or both surfaces being bonded, assembly may take place immediately by means of nothing more sophisticated than manual pressure. However, if assembly is delayed for hours or days, it can then be performed only after provisions have been made for introducing one or more aids to bonding such as solvent reactivation, simple air-drying, heat-drying, or clamping.

Cold-Setting and Hot-Setting Adhesives

Adhesives which cure at room temperature or below are cold-setting, and those that require heat are hot-setting. The distinction is not always apparent, because, for example, the curing rate of any cold-setting adhesive may be increased by introducing heat, but some cold-setting adhesives achieve maximum cure only if cured below room temperature.

Some room temperature curing adhesives consist of two ingredients—the adhesive which often is referred to as the solids content of the formulation and a curing agent. Often this type is a two-part adhesive meaning that the two ingredients must be mixed just prior to use. When mixed, a chemical reaction occurs which begins to set up the cure. As a result, such adhesives have limitations of effective life which may vary from mere minutes to as much as days or a week.

Other types of room temperature curing adhesives are identified as solvent release adhesives. These usually are polymers and/or resins in solution and, after application to surfaces, must await evaporation of the solvent before assembly can be performed, usually with some degree of pressure.

Room temperature curing adhesives may set within an hour at which time bonded parts can be safely handled. But full bond strength is not usually developed until five to seven days. Most of the room temperature curing compositions attain 60 to 85 percent of their ultimate strength within the first 24 hours, after which the curing rate is considerably slower, requiring several days for cure completion.

Hot-setting or heat curing adhesives are available in a variety of forms including solutions, 100 percent solids, and films. The amount and duration of heat required varies, depending upon the chemical nature of the adhesive and the nature of substrates being bonded.

Certain adhesives cure most effectively at intermediate temperatures, in between room temperature and the temperature required for so-called hot-setting adhesives. As a rule, readings higher than room temperature but below 212°F are classified as intermediate.

Classifying Adhesives by Initial Bond

Adhesives often are classified by the method by which each adhesive achieves initial effective bonding: (1) pressure-sensitive, (2) contact, (3) chemical-setting, (4) heat-setting, and (5) liquid-release.

Results from any of the above five methods can be accelerated by using air, heat, pressure, or heat and pressure. There are a variety of means and a considerable number of devices available for providing a proper source of air or heat and the proper type and degree of pressure.

Drying

The drying process removes volatiles from solvent adhesives and eliminates the possibility of solvent entrapment between substrates being bonded. Such entrapment between non-porous surfaces can cause bubbling which, in turn, can weaken good bonds.

Simple exposure to ambient air can be sufficient, particularly if high speed assembly is not required. However, if speed is desirable, the air may be driven onto the exposed, adhesive-coated surface by various means including an electric fan or blower.

To achieve higher drying speeds, force drying with hot air (220° to 240°F) is a common, accepted procedure. A zoned oven in which the air is permitted to circulate is generally preferred for the operation. This ensures that the heat will be distributed evenly across the surfaces. Thermostatic controls are recommended in order to prevent premature setting of thermosetting adhesives. Radiant heat panels are used to dry adhesives.

Curing

Heat or pressure and usually both are required to achieve curing or polymerization of most adhesives. Some adhesives can also be cured at room temperature with little if any pressure, but nearly all can be cured faster by exposure to heat.

Use of Heat

Heat can be furnished by ovens, infrared lamps, dielectric heating, induction heating, electrical resistance, steam, hot water, and oil. The ovens described earlier for drying are most often used for curing when a large number of units can be accommodated at one time. The heat transfer is relatively slow, whether the oven is gas-fired, oil-fired, electrically-heated, or steam-heated. Infrared heating is faster, taking less time to reach a prescribed curing temperature. It is important to note that time measurement should not begin until the oven has reached the temperature prescribed by the adhesive supplier. Often, disastrous results occur simply because the time cycle was measured from the moment the assemblies were placed in the cold oven.

Curing by dielectric heating is sometimes used on assemblies which can absorb energy from a rapidly alternating electromagnetic field. Energy is developed within the dielectric materials by molecular action, a method offering the advantage of fast and uniform heating since heat transfer does not depend on conduction.

Induction heating is similar to dielectric and is suited for metal to metal bonds. It is commonly used for soldering, brazing, and heat treating purposes, but its use for curing adhesives is limited yet.

Electrical resistance as a source of heat for curing necessitates the use of platen presses or special fixtures to be practical.

Steam, hot water, and oil are other heating methods that may be used. Each can provide higher rates of heat transfer than can be provided by a heated oven because liquids heat by conduction rather than convection.

Use of Pressure

Curing usually requires some pressure on the assembly while the adhesive is setting. It helps to establish uniform and complete contact between the adherends and the adhesive, and it helps to establish a uniform thickness of adhesive film.

Regardless of the means used to provide pressure, the method should ensure that the amount of pressure is adequate, that it is retained at a constant level, and that it is distributed uniformly over the work.

Some of the more commonly used pressure means include clamps, bolted fixtures, and various types of spring-loaded jigs, mechanical, pneumatic, and hydraulic presses, pressure chambers such as autoclaves, vacuum equipment, and inflatable bags or tubes.

BONDING WITH HOT MELT ADHESIVES

Bonding or assembly with hot melt adhesives is a distinct technology, and requires applying equipment designed especially for hot melts. Hot melt adhesives, though available in a variety of physical forms, are always 100 percent solid materials that must be subjected to heat to perform as adhesives. The absence of a solvent distinguishes hot melt adhesives from other adhesive types; heat performs the function that solvents perform in conventional wet or liquid type adhesives—to facilitate wetting or penetration of the surfaces to be bonded. Application of heat brings the hot melt material to the liquid state, and it sets on cooling. This feature accounts for the high speed at which a hot melt can produce a bond, undoubtedly its most significant advantage.

In the past the most serious limitation to the use of hot melt adhesives was the inadequacy of available equipment for processing and applying hot melts. Equipment now on the market is quite sophisticated, and the use of hot melts is on the increase, particularly on high speed, high volume packaging lines.

However, there are important variations among the many acceptable types and makes of applying equipment available. For example, the equipment varies in: (1) the manner of feeding and melting the adhesive, and (2) the provisions for exerting the pressure required to carry the molten adhesive to the device which will actually apply hot melt onto a substrate. Of increasing significance, too, is how the hot melt actually is applied—by actual contact of the applying device to the substrate or by so-called jet or spray application.

Spray equipment, used particularly for case sealing, does away with the need of direct contact with the substrate to be bonded. This can be particularly valuable if the hot melt must be applied to a remote or somewhat inaccessible surface area of a packaging form. Since the cases are not usually held to very close dimensional tolerances by suppliers, adhesive application problems are created in case sealing lines using direct contact methods. With a spray device, this problem is eliminated so that cases passing through on the "high" or "low" side nevertheless receive an effective deposit of hot melt.

Also, the spray nozzle escapes any possible contamination from the substrate. Porous substrates can cause a clogging of the nozzle orifice. Spray application seems certain to become the preferred means of making hot melt bonds on corrugated cases and perhaps, on certain types of cartons and other packaging forms.

Two basic types of hot melt applying equipment require contact with the substrate or substrates to be bonded. One is referred to as a nozzle or extrusion applicator; the other, a wheel applicator.

For the bulk of web converting jobs, such as bag making, for which hot melts are suited over the other types of adhesives, the contact type of nozzle application continues to be the most efficient. Also, for a wide variety of carton-closing operations in which very precise amounts and configurations of hot melt must be laid down, it would seem that wheel type applicators will continue to be preferred over spray devices.

The decision to use equipment for spray or contact application of hot-melt adhesives requires a knowledge of both the viscosity and temperature extremes for effectively applying the adhesive. Some applicators of the spray type can handle a viscosity extreme of only 5000 cps and operate at a maximum of 375°F. Other models can handle 10,000 cps at 475°F. Still another type can process hot melts of as much as 15,000 cps and operate at temperatures in excess of 500°F. Contact type applicators vary similarly.

Capacity

Applying equipment varies in the amount of properly heated adhesive per unit of time it can apply. Regardless of type, every hot melt applicator is identified by the maker in terms of melting capacity. Some will provide the amount of grams or pounds of the adhesive the unit can melt and apply per minute, and also how much coverage it can furnish per minute. Coverage is expressed in terms of the width and thickness of the adhesive seam per unit of time.

It is essential to understand that the output capacity is unrealistic unless it has been measured with an awareness of the permissible differential in temperature of the adhesive from the time it is subjected to heat to the time when it is actually emerging from the nozzle or wheel device. It may be true to state that the capacity of the applying equipment is, for example, 50 gm/min if the temperature at which the hot melt adhesive must be actually dispensed to a substrate is only 250°F. However if the desired temperature is 350°F, it may no longer be possible for the same unit to dispense as much as 50 gm/min of adhesive within the permissible application temperature range for the formulation. The capacity rating of the unit, for this new requirement will, therefore, be somewhat smaller.

Other Factors

Another point about hot melt equipment is that it can apply adhesives upward as well as downward, horizontally, or in a combination of these positions. This flexibility can be of utmost significance to an end user whose parent machinery is so constructed that the application of the hot melt must be done in a particular direction.

Power requirements for the various types and models of equipment do vary and are provided by the manufacturers. Heat is a vital necessity in any case, and heaters powered by electricity are required. The number of heaters and wattage of heaters vary. Cost of electric power is important to the economics of usage and depends on the prevailing power costs at the using plant.

Prices for hot melt equipment range from $800 to $5000 and will require some maintenance and servicing, as does most machinery. Training of in-plant personnel to perform such maintenance and servicing can be most helpful. The advantages that hot melt systems can provide in terms of production speed-up can be extremely significant and quickly pay off the capital investment required.

chapter 6

inspection and quality control

Quality control of a bonded joint involves making sure that a certain level of reliability is achieved in a joint. Quality control also performs the important function of compiling statistics on defects for the availability of all accountable bonding personnel. Reliability is achieved by a thorough in-process control from the time the raw materials come in until the last finishing operation is completed.

PROCESS CONTROL

Process control starts with the receipt of the raw materials which are, in this case, the adhesives, catalysts, etc. The purchase order normally defines the required quality properties of the material by virtue of an actual statement of requirements or by what is called out on the Material Specification. The inspection requirements are normally specified in the Material Specification as Quality Acceptance Tests or as Receiving Inspection Requirements.

Container Condition

The first of these requirements normally is the condition of the container. The following things should be looked for when inspecting the condition of the container:

1) Damage (physical) to a container of film adhesive can rupture its sealed wrapper allowing moisture, dirt, etc. to reach the adhesive rendering it contaminated. Damage can render a pail of liquid measure unusable in automatic dispensing equipment.
2) Leakage of liquid adhesive can change the ratio of the catalyst to the base resin if pre-measured kits are involved. It can also result in the receipt of less material than purchaser needs and is paying for.
3) Identification of a container needs to include: product number, manufacturer's name, manufacture date, batch or lot number, shelf life, storage conditions, manufacturer's instructions for use, and safety precautions.

After quality control inspects the adhesives upon delivery, production can be assured that the incoming materials meet specifications.

Before the adhesives are issued to manufacturing, quality control must verify their physical and chemical properties. These are usually included as Receiving Inspection requirements. Inspection consists in looking for:

1) Strength
2) Resistance to environment
3) Thickness
4) Weight
5) Color (inspection aid)
6) Slump
7) Adhesion
8) Percent solids
9) Ingredients
10) Flow
11) Tack
12) Viscosity
13) Elongation
14) Pot life

The tests to be required usually are those which will most quickly and accurately indicate deficiencies for the properties needed to support the designed use.

Production of adhesive bonded parts involves: (1) prefit, (2) surface preparation, (3) adhesive application, (4) assembly, and (5) cure.

Prefit

All detail parts must be *dry* fitted together to insure a close contact of the faying surfaces. If two or more detail parts do not fit prior to being bonded, they are not likely to fit well enough after bonding to produce a good joint. If a high production rate exists where a reproducible fit accuracy can be established, then prefit can sometimes be omitted. First article fits can be checked by using tool proofing films which produce an imprint or image of the joint fit. This can greatly reduce the risk factor of poor fit where expensive components are involved. After prefit conditions are verified, each detail part fitted in that assembly should be identified as such to facilitate mating of those specific parts after adhesive application.

Process control test panels or excess *tag end* portions of the assembly are included with the *kit* of prefitted details at this point and are verified at the time of prefit inspection. These process control test specimens must be processed through all operations simultaneously as the end product. They will be tested after curing to verify the adhesive batch, surface preparation, and other processing conditions used on that end item.

Surface Preparation

The appropriate surface preparation as described in Chapter 4 must be monitored for proper sequence, solution concentration, solution temperature, solution contaminants, etc. The most commonly used check for proper condition prior to drying after surface treatment is a water break free test. After final rinse, the surface is visually checked for the presence of a continuous film of water without any breaks in the water film. From this point on, the processing of treated parts must be carefully monitored to insure maintenance of clean substrate surface. In general, most surface conditions for optimum bonding of adhesives necessitate that the prepared surfaces be coated with adhesive primers or the adhesive itself within a prescribed period of time as designated in Table VI–1.

Table VI–1. Time Interval Limit for Adhesive Coating of Prepared Surfaces.

Type of Surface Preparation	*Time Interval (hr)*
Clad aluminum/etched	Within 16
Bare aluminum/etched	Within 8
Anodized aluminum	Within 24
Steel/etched	Within 8
Titanium/conversion coated	Within 16

Adhesive Application

Most structural film adhesives require use of a primer to protect prepared surfaces, resist peel loads, and promote wetting of the substrate surfaces by the adhesive when low flow adhesives are involved. Adhesive primers are usually spray applied by air or by airless spray systems. Roller or brush is sometimes used on small areas or when spray equipment is not available. The primer coat must be air dried and sometimes oven-baked to remove solvents. The thickness of the prime coat usually will affect the adhesive bond strength and must be controlled and verified. This is usually done by periodically certifying the prime applicator, and monitoring part primer thickness after drying.

Film adhesives are applied by removing a paper or plastic separator/protective film and laying the adhesive on the faying surface smoothly, taking care not to allow wrinkles to occur nor air to become entrapped between the adhesive and the substrate surface. A common workmanship error is made by not removing the separator film before assembly of the detail parts. Some companies utilize special inspection check-off points to insure separator film removal.

The batch number, lot number, time, and date of application and adhesive type should be logged into the inspection record for traceability should a failure occur. The shop life expiration date and time should also be logged to aid in controlling assembly and cure of adhesive coated parts within the prescribed working life of the adhesive.

Assembly

The adhesive coated detail parts are usually joined in a tool or holding fixture. Cleanliness and proper preparation of the tools should be verified. Time limits on the surface preparation, shop life of the adhesive, and remaining time during which the adhesive must be cured needs verification at the point of assembly.

Assembly of detail parts in their proper sequence and fit should be verified. Maintenance of cleanliness and atmosphere control is important. The atmosphere to which the parts and the adhesive are exposed must be controlled from the time the detail parts are prepared for adhesive application until the cure is initiated. Controlling the atmosphere normally involves: (1) keeping the temperature between 65° and 90°F, (2) keeping relative humidity between 20 and 65 percent, (3) filtering of all incoming air to preclude airborne contaminants, and (4) maintaining of a slight positive pressure differential between the controlled environmental area and all surrounding areas. Temperature and humidity indicators of the recording type have become minimal environment verification equipment.

Curing

Curing an adhesive in any bond joint usually is a time-temperature-pressure function. Regardless of the means of controlling each of these three variables, the documentation-verification means are basically the same.

Controlling the length of cure time can be by manual or automatic timing devices. Verification is usually documented on a cure chart taken from a temperature and/or pressure recorder; recordings of pressure and temperature are obtained in the same manner.

The facility used for applying heat must be certified for its basic capabilities and uniformity with respect to its intended use. Points that must be considered are: (1) heat-up rate, (2) maximum temperature limits, (3) temperature range or spread during heat-up and at cure temperature, and (4) cool down characteristics. The certification should make the facts clear about the curing facility. The same degree of verification (namely certification) is needed for the pressure characteristics of the facility if it is an autoclave, a vacuum system, or a press.

BOND INSPECTION AND TESTING

Innumerable procedures have been developed during the past two decades to inspect and determine the quality of adhesive-bonded joints. Inspection and testing are conducted for two major reasons: (1) to evaluate the quality or performance of adhesive-bonded joints under specified loading and environmental conditions, and (2) to establish quality control procedures on the basis of the test results. However, the multiplicity of inspection and testing procedures can produce considerable confusion when the results are analyzed. For example, the lap shear joint is probably the most widely accepted design to be used in evaluating adhesives, adherend surface preparation, bonding procedures, etc. However, this joint design frequently bears little resemblance to the joint that will be used in a structural application. In addition, other tests may be used to evaluate joint quality, and a specific joint may pass one test but not the other. In such cases, caution must be exercised in comparing the quality of a number of joints. Usually, the tests are weighted, and an average quality level is obtained. With all tests there is scatter among the test data. Data obtained from tests of adhesive-bonded joints generally contain more scatter than comparable joints made by welding or brazing. For this reason a statistical approach in evaluating adhesive-bonded joints is strongly recommended.

The selection of methods for inspection and evaluation of joint quality depends on factors such as: (1) the nature of the adherends or substrates, (2) the type of adhesive, (3) the joint design, and (4) the service conditions. Methods that are suitable for use with adhesive-bonded metallic substrates are not necessarily useful when nonmetallic substrates are joined. Similarly, tests used to evaluate the performance of simple lap joints are of no use in evaluating the soundness of adhesive-bonded sandwich structure. The usefulness and limitations of inspection and testing methods must be clearly understood before they can be successfully employed. Fortunately, the industry has progressed to the stage that inspection and test requirements are quite well detailed for specific joint systems.

The provisions of inspection and test methods for adhesive-bonded joints are detailed in specifications issued by Government agencies, by technical societies, and by the industrial firms engaged in joining materials with adhesives. These procedures evolved over a period of time and reflect the thinking of many individuals; they are subject to change as new and improved techniques are developed for evaluating adhesive-bonded joints. It is beyond the scope of this chapter to review all of the available inspection and test methods. Instead, the most widely used methods are emphasized. Most of the methods discussed are covered by ASTM (American Society for Testing and Materials) specifications.

Nondestructive Testing of the Bond

Appearance. Visual inspection can be used to a limited degree to determine the quality of adhesive-bonded joints. The usefulness of visual inspection is limited by the joint design, the type of adhesive, and the nature of the substrates. Considerable information can be obtained if the substrates are transparent or translucent. In such cases a dye can be introduced into the adhesive and the bonded area can be illuminated by a strong light. Defects such as unfilled areas, voids, bubbles, cracks, etc., can be detected by the transmitted light. The exterior appearance of the adhesive-bonded joint can yield a limited amount of information on bond quality. Unfilled areas and voids can sometimes be detected by visual inspection; misaligned joints are readily visible. The appearance of the flash along the joint and its texture are clues to the effectiveness of the curing process. The flash should be well filleted to the substrates.

Tap Test. One of the first methods used to evaluate the soundness of an adhesive-bonded structure was the tap test, in which a coin is used as a special tapping hammer (1). An experienced inspector taps the structure lightly and analyzes the resulting tone; tone differences indicate inconsistencies in the bonded joint. Of course, the success of the tap test depends on the skill of the inspector, the noise level, and the type of structure, etc. Some improvement in the tap test can be achieved by using a solenoid-operated hammer and a microphone pickup. The resulting electrical signal can be analyzed on the basis of amplitude and frequency.

Ultrasonic Tests. The success of the tap test, although limited, led to the use of ultrasonics to determine bond quality. Ultrasonic energy can be introduced into the adhesive bonded structure by several methods whose choice is dependent on the type of structure being inspected (1). Short pulses of ultrasonic energy can be introduced on one side of the structure and detected on the other side; a void will prevent the ultrasonic energy from passing through the structure. This method has been used successfully with metal-to-metal and metal-to-core bonds, fiber glass-reinforced plastic bonds, phenolic-to-phenolic and phenolic-to-core bonds, and similar joints. In the pulse-echo method, short pulses of ultrasonic energy are introduced into the structure; these pulses are reflected by the bond line or by the far surface of the bonded assembly producing the echo which is analyzed. This method has been most successful with metal-to-metal and metal-to-core joints where the presence of a large echo or loss of back surface echo indicates an unbonded area.

Shear waves or Lamb waves can also be introduced into the structure with a wedge-shaped transducer. Again, these waves are reflected and analyzed. This technique is very successful with adhesive-bonded sandwich structure.

The sweep-frequency resonance method of ultrasonic inspection has the advantage of producing an estimate of bond strength in metal-to-metal and metal-to-core structures as well as similar structures made from nonmetallic materials. The energy that is introduced into the structure is varied over a wide range of frequencies; the resonance set up by the probe, face sheet, adhesive, and the remainder of the structure is observed. One of the best known and most reliable ultrasonic testing systems for inspecting adhesive-bonded structures in the laboratory is based on the sweep-frequency resonance method; it is known as the Fokker Bond Tester (1) and is used by some aircraft firms. Some aircraft and helicopter manufacturers also use high frequency pulse echo facsimile recordings to test in-service aircraft or completed aerospace vehicles where the only available surface for testing is often the outer mold line of the airframe.

Radiography. Radiographic inspection techniques have proven to be successful for the detection of defects in adhesive-bonded metal-to-metal joints and metal sandwich structures (2). In the case of metal-to-metal joints, the adhesive must contain some metal powder or some other suitable filler to create enough contrast to make defects visible; the same procedure could be used with nonmetallic adhesive-bonded joints. An experienced inspector, using radiography, can often detect undesirable concentrations of adhesive or evaluate the quality of adhesive-bonded honeycomb sandwich structure. A damaged core is readily detectable by radiographic means. Damage can occur in handling; or it may be the result of unequal pressure during the bonding cycle.

Proof Tests and Leak Tests. The inspection methods discussed so far can be used to detect flaws and defects in adhesive-bonded joints; however, they cannot produce quantitative data on the stress level that might induce mechanical failure of the joint. If a high degree of reliability is required, it is necessary to proof test a portion (or all) production units (2). Specific proof tests cannot be recommended to cover all situations, but some general principles can be advanced. The proof test should simulate actual service conditions in the manner in which the joint or structure is loaded; the stress level should be higher than that expected in service, and the duration of the test should reflect the expected life of the joint. The environmental conditions (heat, moisture, atmosphere, etc.) to be encountered during service should also be considered in proof-testing. If a closed structure is produced by adhesive-bonding techniques, the structure can be tested for leaks by various methods such as hydrostatic tests and helium leak tests.

Destructive Testing of the Bond

Numerous test procedures have been developed to evaluate the properties of adhesive-bonded joints and structures; most of these methods have been developed under the auspices of Committee D-14 of the American Society for

Testing and Materials and appear in the *ASTM Book of Standards*, Part 16 (3). This publication is revised on a yearly basis, and the latest edition should be consulted for up-to-date information.

Considerable care must be exercised in the accumulation and reporting of data on the mechanical testing of adhesive bonds. These data should include: (1) the number of specimens tested, (2) complete identification of the specimens, (3) method of specimen preparation, (4) type of test, (5) testing conditions, (6) specimen size and configuration dimensions, and (7) strength data.

Tensile Tests. The relative tensile strength of adhesives can be determined with bar- and rod-shaped specimens in accordance with the provisions of ASTM D2095-62T. The test specimen and attachment fixtures are shown in Fig. 6–1.

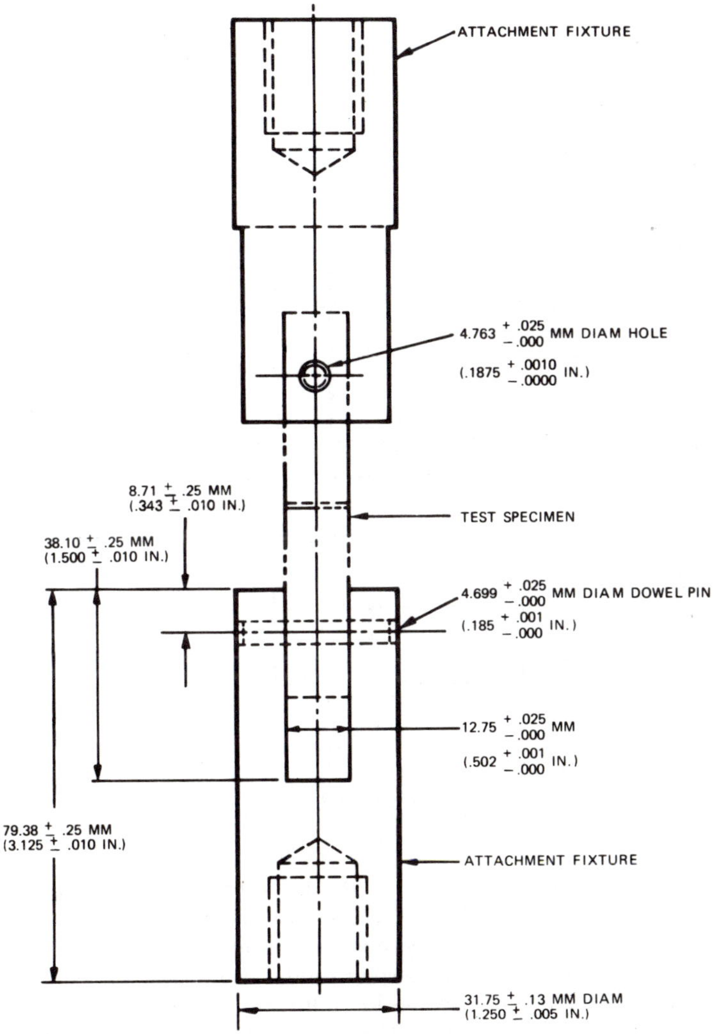

Fig. 6–1. Test specimen and attachment fixtures (from ASTM D2095-62T).

This test was developed by the U. S. Naval Ordnance Laboratory and is particularly useful for adhesives that do not require too much pressure during the curing cycle. The specimens are prepared in accordance with ASTM D2094-62T. A minimum of five specimens are tested for each test condition. All specimens except those in which both substrates are metal should be conditioned prior to testing for 40 hours at a temperature of 73.4° ± 2°F (23° ± 1°C) and a relative humidity of 50 ± 2 percent; metal-to-metal joints can be tested as soon as the specimen reaches an equilibrium temperature of 73.4° ± 2°F (23° ± 1°C). Testing is conducted in a standard tensile testing machine; the rate of loading is 2400 to 2800 psi of bond area/min. The maximum load at which failure occurs is recorded, and the percentage cohesion failure, adhesion failure, and substrate failure are estimated by visual inspection. The tensile strength is calculated in psi.

The cross-lap specimen (Fig. 6–2) is used to determine the relative tensile strength of adhesives; this specimen has the advantage of fewer limitations on the substrates. The test, ASTM D1344-57 (1965), was originally developed

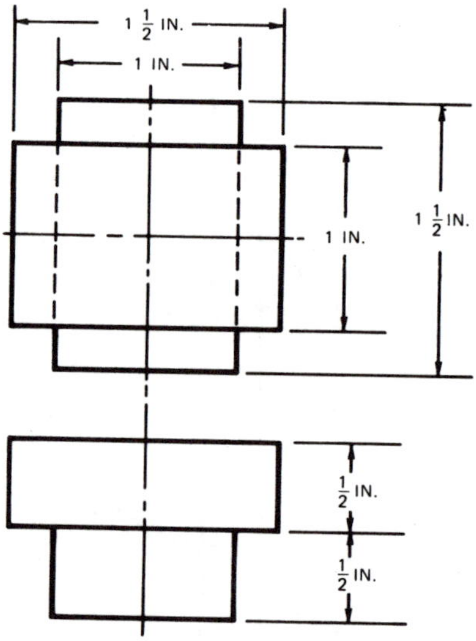

Fig. 6–2. Cross-lap specimen (from ASTM D1344-57).*

*Permissible variation for all three dimensions of the glass test specimen may be 1/64 in.

to evaluate the adhesion of glass to itself and other materials. During testing, the specimen is held in self-aligning grips. At least 10 specimens for each adhesive must be tested. Glass-to-glass and glass-to-metal joints need not be preconditioned before testing; glass-to-wood should be conditioned at 73.4° ± 2°F (23° ± 1°C) and 50 ± 2 percent relative humidity for seven days or until they have reached equilibrium. Testing should be done at a loading rate of 600 to 700 lb/min. The same data are recorded as indicated above.

The relative tensile strength of wood-to-wood and metal-to-metal joints can be determined with specimens cut to the dimensions shown in Fig. 6–3. The test grips for the specimens are shown in Fig. 6–4 (ASTM D894-49 (1965). The test procedures outlined by this standard are very similar to those discussed immediately above.

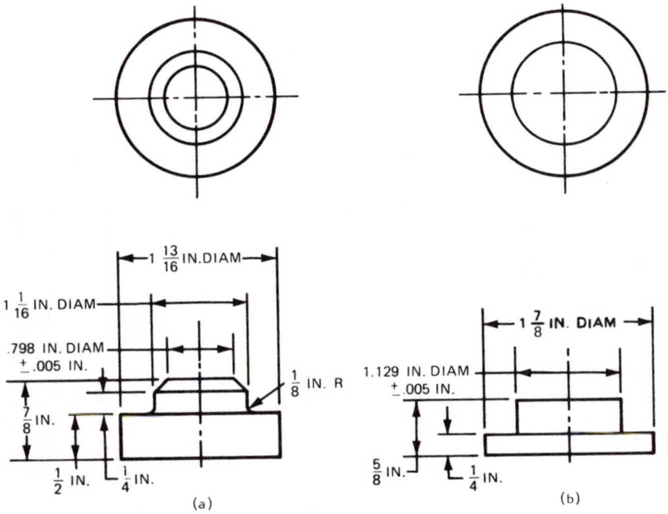

Fig. 6–3. Test specimens: *a* wood specimen (hard maple), *b* metal specimen (from ASTM D897).*

*Surfaces of metal specimens shall be ground flat and parallel. It is preferable to grind all metal specimens at the same time.

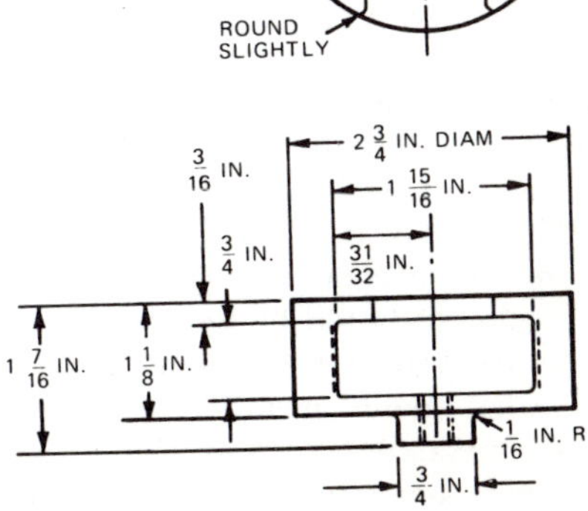

Fig. 6–4. Test grips, cold rolled steel (from ASTM D897).

Shear and Lap Shear Tests. A test to determine the relative shear strength of adhesive-bonded joints that are loaded in compression are discussed in ASTM D905-49 (1965). This method is primarily intended to evaluate adhesives for wood. The testing machine should have a capacity of at least 15,000 pounds in compression and should be equipped with a shearing tool that has a self-aligning seat (see Fig. 6–5). The form and dimensions of the test specimen are shown in Fig. 6–6. Before testing, the adhesive-bonded specimens should be conditioned at 73.4° ± 2°F (23° ± 1°C) and a relative humidity of 50 ± 2 percent for seven days or until the specimens reach equilibrium. Testing should be done using a loading rate of .015 in/min. Data to be recorded and/or calculated include specimen identification, conditioning procedure, test conditions, number of specimens tested, and maximum, average, and minimum shear strength.

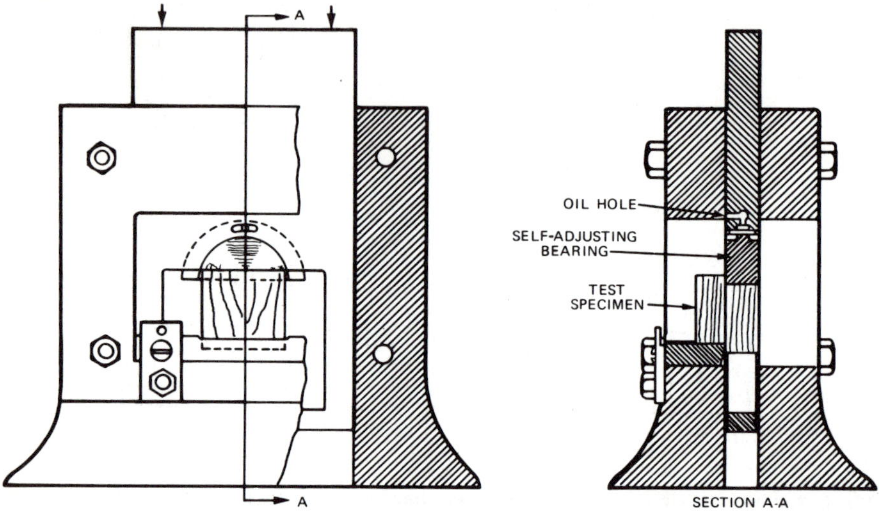

Fig. 6–5. Shearing tool (from ASTM D905-49 (1965)).

A somewhat similar test has been developed to determine the relative shear strength of adhesive-bonded metal-to-metal joints by compression loading; it is designated ASTM 2182-63T. The testing machine should have a capacity of at least 10,000 lb and should be capable of maintaining a uniform loading rate of 1200 to 1400 psi/min. The test specimen has the form and dimensions shown in Fig. 6–7. The shearing tool is provided with a heating coil so that tests can be made at room and at elevated temperature. The specimen should be placed in the shearing tool and loaded in compression at a rate of 1200 to 1400 psi/min until failure occurs. For elevated-temperature testing, the adhesive-bonded specimen should be heated to the required temperature and held for 10 min before testing; the specimen can be heated in the shearing tool. The data recorded should include the specimen identification, bonding conditions, bond thickness, test temperature, conditioning treatment, number of specimens tested, shear stress at failure, and nature of failure.

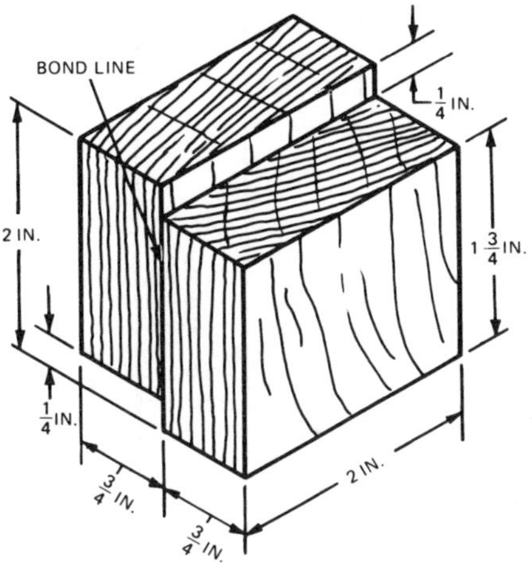

Fig. 6–6. Test specimen (from ASTM D905-49 (1965)).

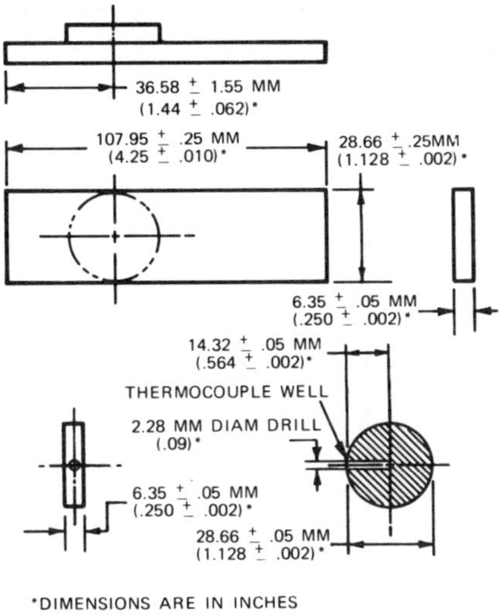

Fig. 6–7. Test specimen (from ASTM D2182-63T).

ASTM D1002-64 is a widely used standard for determining the relative shear strength of adhesive-bonded metal-to-metal specimens when loaded in tension. The testing machine should have a capacity such that the breaking load of the specimens falls between 15 and 85 percent of the full-scale capacity. The machine should be capable of maintaining a loading rate of 1200 to 1400 psi/min. The specimens should have the form and dimensions shown in Fig. 6–8, and should be cut from test panels such as those shown in Figs. 6–9 or 6–10. The test panels should be free from burrs or bevels and should be smooth before cleaning and bonding; the adhesive should be applied to the test panels in such a manner that it occupies a space about ¼ in. wider than the specified overlap. The individual specimens are placed in the self-aligning grips of the testing machine and loaded in tension at a rate of 1200 to 1400 psi/min until failure occurs. The data to be recorded are similar to those listed under ASTM D2182-63T.

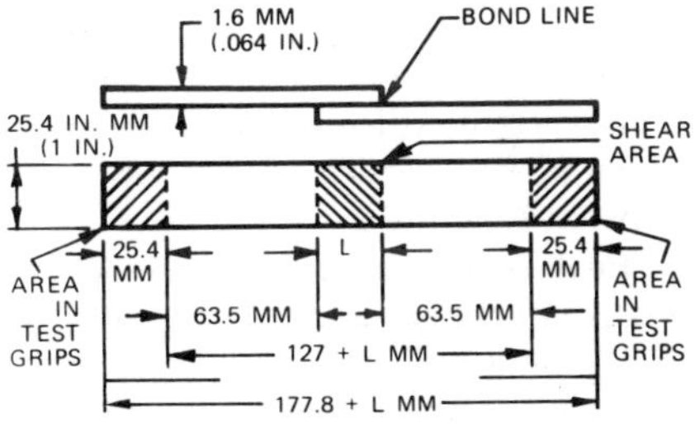

Fig. 6–8. Lap shear specimen (from ASTM D1002–64).

Other standards for determining the shear properties of adhesive-bonded joints include ASTM 906-64 (Strength Properties of Adhesives in Plywood Type

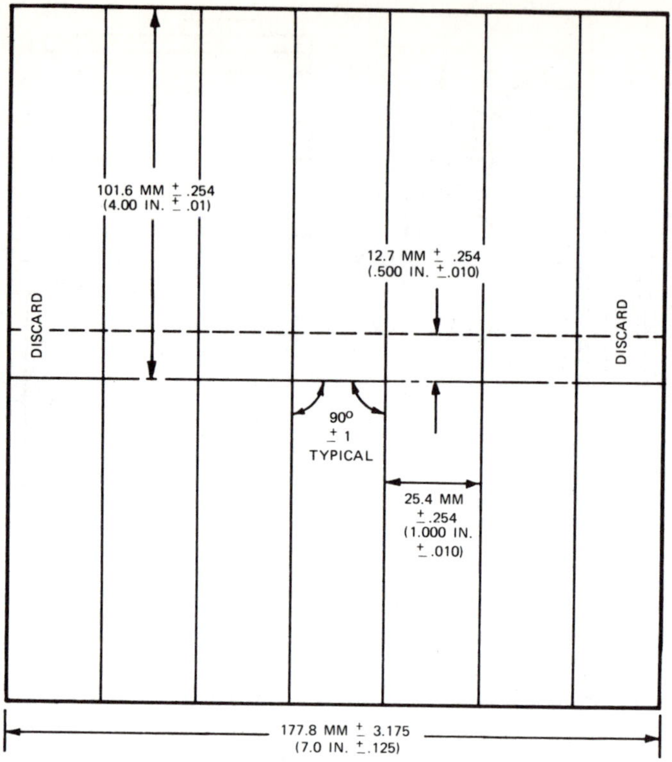

Fig. 6–9. Standard test panel (from ASTM D1002-64).

Construction in Shear by Tension Loading), ASTM D2293-64T (Creep Properties of Adhesives in Shear by Compression Loading), ASTM D2294-64T (Creep Properties of Adhesives in Shear by Tension Loading), and ASTM D2339-65T (Strength Properties of Adhesives in Two-Ply Wood Construction in Shear by Tension Loading).

Peel Tests. The T-peel test (ASTM D1876-61T) is designed to determine the relative peel resistance of adhesive bonds between flexible adherends. The form and dimensions of the test specimen and the test panel from which individual specimens are cut are shown in Fig. 6–11. The testing machine should be so selected that the maximum load on the specimens falls between 15 and 85 percent of the upper limit of the loading range; the rate of movement should remain constant under fluctuating loads. A chart should be produced during testing with Inches of Separation as one coordinate and the Applied Load as the other coordinate. The test panels should be bonded in accordance with accepted practice; at least 10 specimens should be tested for each adhesive. Before testing, the nonmetallic specimens should be conditioned for seven days at 73.4° ± 2°F (23° ± 1°C) and 50 ± 2 percent relative humidity. The peel test should be conducted at a constant head speed of 10 in/min. Data to be recorded include specimen identification, bonding conditions, conditioning treatment, test conditions, type of failure, and peel strength (average, maximum, and minimum).

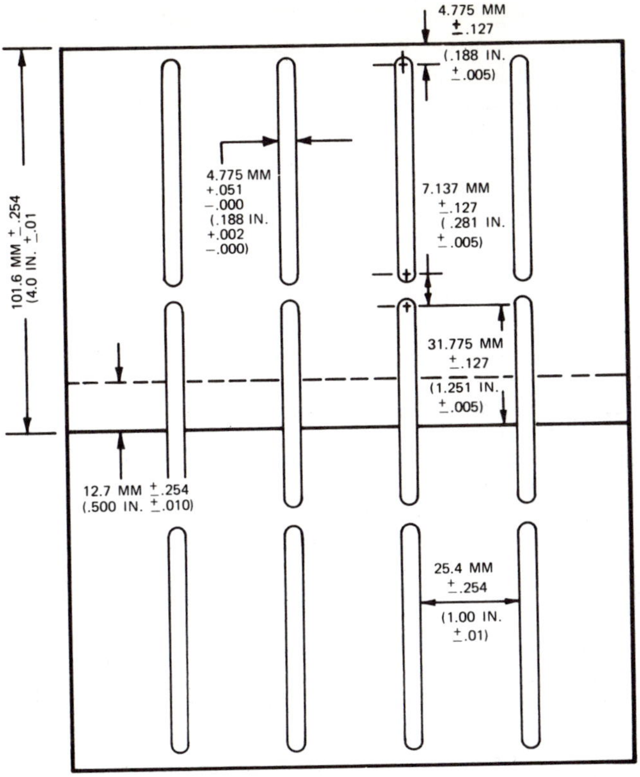

Fig. 6–10. Optional panel for acceptance tests (from ASTM D1002-64).

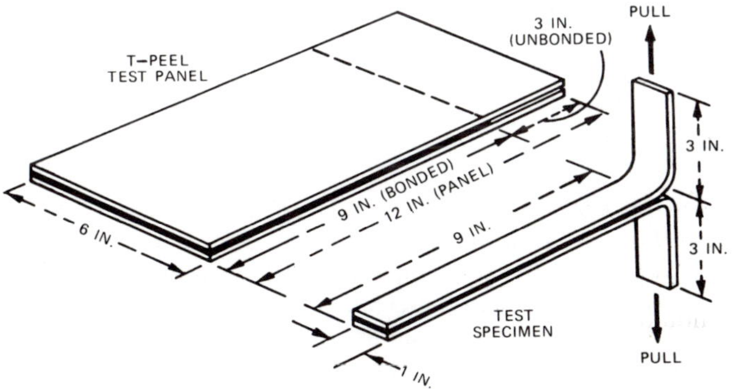

Fig. 6–11. Test panel and T-peel specimen (from ASTM D1876-61T).

The climbing drum peel test (ASTM D1781-62) was developed by the U. S. Forest Products Laboratory and is used to determine the relative peel resistance between a relatively flexible adherend and a rigid adherend. The peeling apparatus is shown in Figs. 6–12, 6–13, and 6–14 together with a section of sandwich structure. This equipment can be also used to test laminated structures. The testing machine to which the peeling apparatus is attached should be so selected that the maximum load on the specimen falls between 15 and 85 percent of the machine capacity. The rate of movement should be constant under fluctuating loads. For laminated assemblies, the specimen should be 1 in. wide and at least 10 in. long; sandwich structure specimens should be 3 in. wide and at least 12 in. long. The specimen should be conditioned for seven days (unless

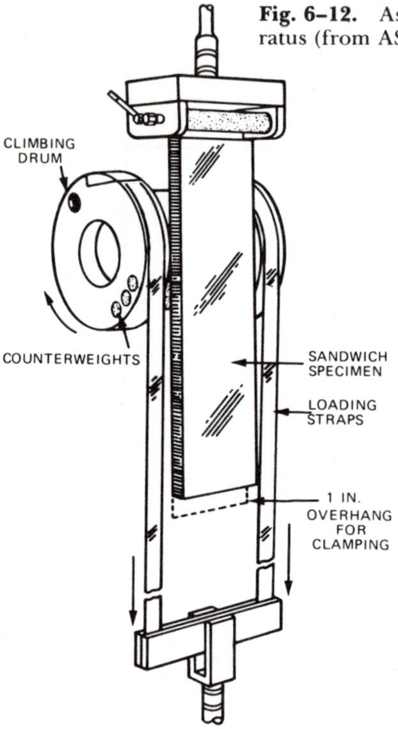

Fig. 6–12. Assembly of peeling apparatus (from ASTM D1781-62).

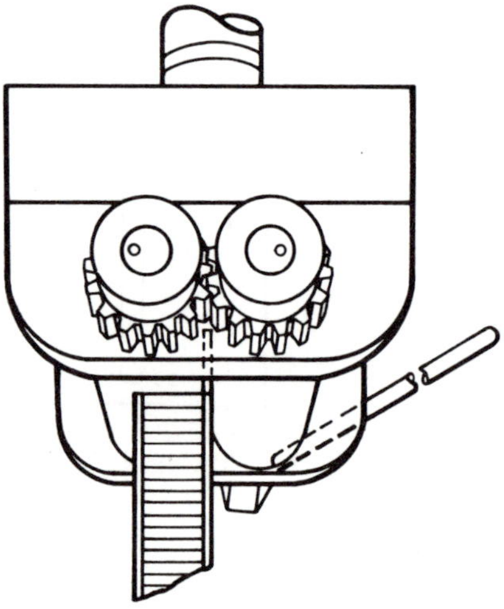

Fig. 6–13. Top clamp (from ASTM D1781-62).

a shorter period is adequate) at 73.4° ± 2°F (23° ± 1°C) and 50 ± 2 percent relative humidity. The peel resistance should be determined over at least 6 in. of the bonded area by loading the peel apparatus at a cross-head rate of 1.00 ± 0.10 in/min. Data on load vs distance peeled data should be recorded on a chart. Data to be recorded include specimen identification, bonding conditions, conditioning procedure, machine calibration procedure, number of specimens, dimensions of specimens, average peel torque, and method of failure.

For rubbery adhesives, a strip-back peel test is available and is discussed in ASTM D903-49 (1965).

Cleavage Test. ASTM D1062-51 (1961) discusses a method of determining the comparative cleavage strength of adhesive-bonded metal-to-metal joints under tension loading using the test specimen shown in Fig. 6–15. The test specimens should be prepared in accordance with recommended procedures;

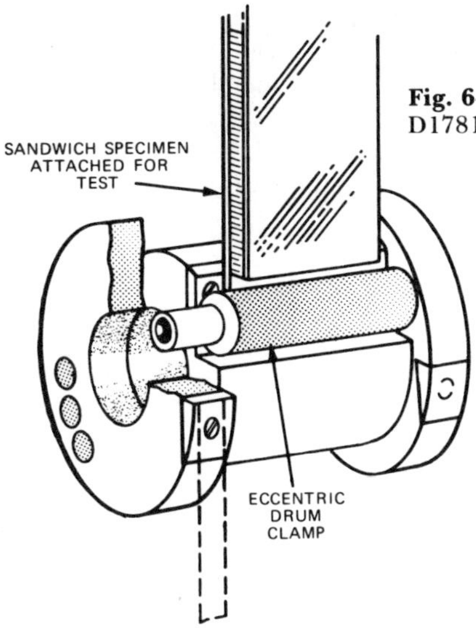

Fig. 6–14. Drum clamp (from ASTM D1781-62).

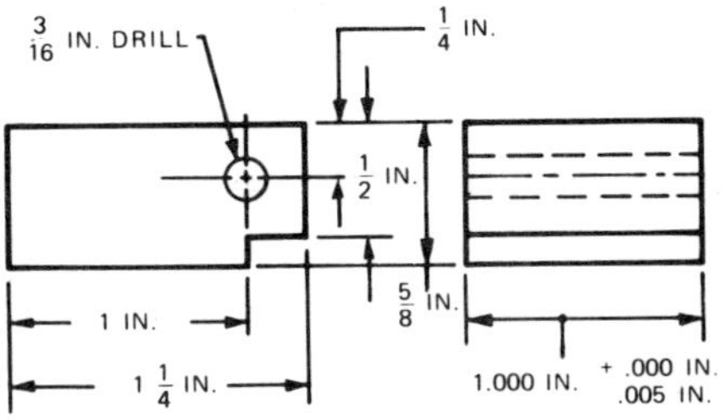

Fig. 6–15. Cleavage test specimen (from ASTM D1062-51 (1961)).

no conditioning treatment is required before testing. The specimens should be attached to the testing machine with self-aligning grips (Fig. 6–16) and loaded at a rate of 600 to 700 lb/min to failure. Included among the data to be reported are specimen identification, bonding procedure, testing conditions, number of specimens tested, and cleavage strength (maximum, minimum, and average).

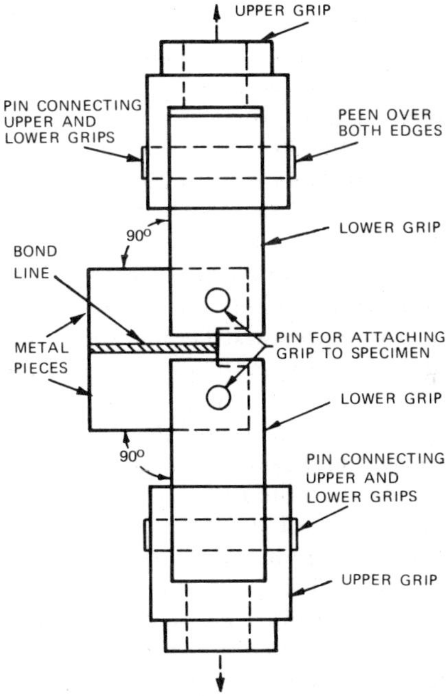

Fig. 6–16. Assembly of grips and test specimen (from ASTM D1062-51 (1961)).

Impact. The comparative impact strength of structural adhesives in shear can be determined in accordance with ASTM test method D950-54 (1961). This method can be used to evaluate the impact properties of metal-to-metal joints, wood-to-wood joints, joints between laminar materials, joints between dissimilar materials, etc. The test specimen should have the form and dimensions shown in Fig. 6–17. The specimen is held in a grip such as that shown in Fig. 6–18; the

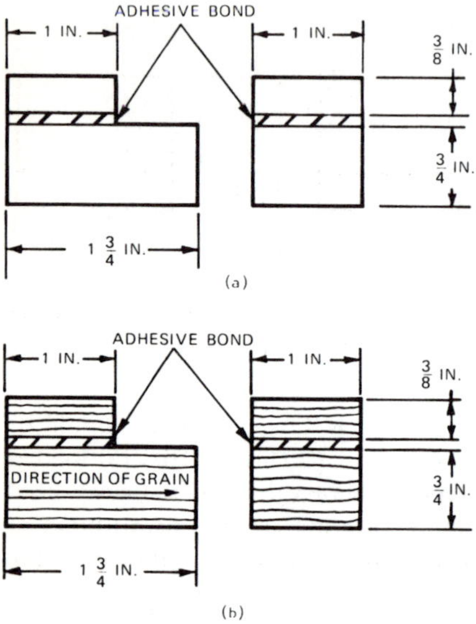

Fig. 6–17. Impact test specimens: *a* metal to metal, *b* wood to wood (from ASTM D950-54 (1961)).

grip is mounted in a standard impact machine. The test specimens are bonded according to accepted practice. Preconditioning is not required for metal-to-metal joints; wood-to-wood joints should be conditioned for seven days at 73.4°±2°F (23°±1°C) and 50 ±2 percent relative humidity. In the case of metal-to-metal joints, at least 10 specimens should be tested; at least 20 wood-to-wood joints should be tested. Testing should be conducted with a head velocity of 11 ft/sec. Data to be recorded include specimen identification, bonding conditions, conditioning procedure, testing conditions, number of specimens tested, and maximum, minimum, and average impact values.

Environmental Temperature and Humidity Test. The effect of temperature and humidity on the properties of adhesive-bonded joints can be determined in accordance with the provisions of ASTM D1151-61. This test specifies conditions for determining the performance of adhesive bonds when subjected to continuous exposure at specified conditions of moisture and temperature. This standard method of test is to be used in conjunction with one of the following methods: ASTM D897 (Tensile Properties of Adhesives), ASTM D903 (Test for Peel or Stripping Strength of Adhesives), ASTM D906 (Properties of Adhesives in Plywood Type Construction in Shear by Tensile Loading), ASTM D1002 (Strength Properties of Adhesives in Shear by Tension Loading), or any other ASTM method for determining the strength of adhesive-bonded joints. The test specimens should have the form and dimensions specified in the particular

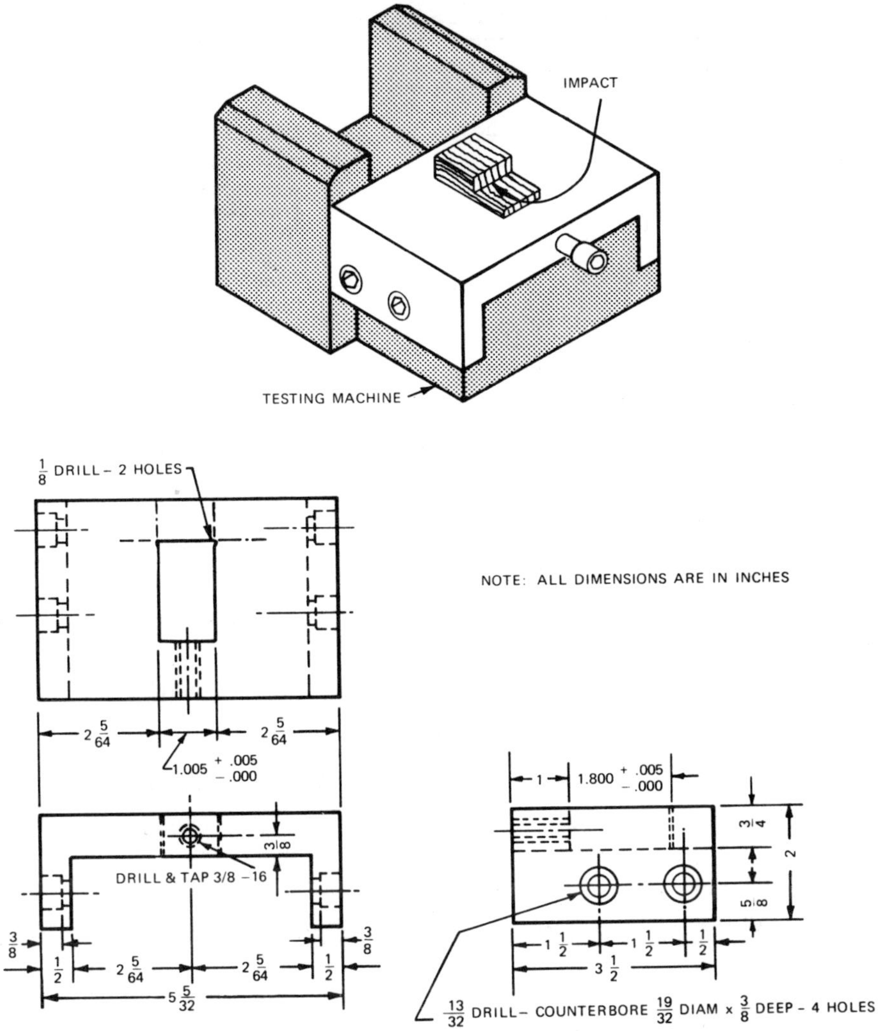

Fig. 6–18. Impact machine jig (from ASTM D950-54 (1961)).

test method being used. The standard test exposures are listed in Table VI–2. Matched specimens should be selected for control and exposure treatment. All specimens should be conditioned prior to test for seven days at 73.4° ± 2°F (23° ± 1°C) and 50 ± 2 percent relative humidity. The control specimens should be tested immediately after conditioning in accordance with the provisions of the test method being used. The values obtained are averaged and recorded as the original strength. The remaining specimens should be subjected to the designated exposure conditions for the length of time specified in the test method. These specimens should be tested after exposure, or tested after a reconditioning cycle of the type discussed above. These values are averaged and

Table VI–2. Standard Test Exposures (D1151) (3).

Test Exposure Number	*Temperature**		*Moisture Conditions*
	°C	*°F*	
1	−57	−70	As conditioned
2	−34	−30	As conditioned
3	−34	−30	Presoaked†
4	0	32	As conditioned
5	23	73.4	50 percent relative humidity
6	23	73.4	Immersed in water
7	38	100	88 percent relative humidity
8	63	145	Oven, uncontrolled humidity
9	63	145	Over water‡
10	63	145	Immersed in water
11	70	158	Oven, uncontrolled humidity
12	70	158	Over water‡
13	82	180	Oven, uncontrolled humidity
14	82	180	Over water‡
15	100	212	Oven, uncontrolled humidity
16	100	212	Immersed in water
17	105	221	Oven, uncontrolled humidity
18	149	300	Oven, uncontrolled humidity
19	204	400	Oven, uncontrolled humidity
20	260	500	Oven, uncontrolled humidity
21	316	600	Oven, uncontrolled humidity

*The tolerance for test temperature shall be ±1°C or 1.8°F up to 82°C or 180°F, and ±1 percent for temperatures above 82°C or 180°F.

†Presoaking shall consist of submerging specimens in water and applying vacuum at 20 in. of mercury until weight equilibrium is reached.

‡The relative humidity will ordinarily be 95 to 100 percent.

recorded as the strength after exposure. The performance of the joint is calculated by dividing the strength after exposure by the original strength and multiplying the result by 100. Data to be recorded include the specimen identification, bonding conditions, description of test specimens, designation of ASTM test method, test conditions, number of specimens tested, and performance.

Method ASTM D2182 can be used for determining the shear strength of adhesive-bonded metal-to-metal joints at elevated temperatures. This method has been discussed earlier.

Cryogenic testing of adhesive-bonded metal-to-metal joints can be conducted in accordance with one of the ASTM test methods, if a cryostat is constructed to surround the test specimen.

Environmental Corrosion Test. The resistance of adhesive-bonded joints to various corrosive media can be determined by ASTM test method D896-66. This method merely outlines the manner in which the test specimens are conditioned before testing. The actual test procedures are governed by the method selected for joint evaluation. That is, the effect of corrosive materials on the tensile, shear, peel, and cleavage properties of adhesive-bonded joints can be determined by the use of the appropriate test method and ASTM D896. Chemical reagents should be selected from ASTM D543 (Test for Resistance of Plastics to Chemical Reagents), and oils and fuels should be selected from ASTM D471 (Test for Change in Properties of Elastomeric Vulcanizates Resulting From Immersion in Liquids). Tests can be conducted using other materials such as jet fuels, hydraulic fluids, liquid coolants, etc., if required. Matched test specimens should be prepared for this test; half of the specimens should be tested in the as-bonded condition, and the remainder should be tested after exposure in the corrosive reagent. The latter specimens should be immersed in the selected reagent for seven days at 73.4° ± 2°F (23° ± 1°C). Closed chambers should be used to minimize evaporation and/or change in concentration of the reagent. The control specimens and the treated specimens should then be tested at room temperature in accordance with the selected test method. The percentage change in average strength resulting from the immersion treatment should be recorded along with the selected ASTM test method, the type of specimens, the bonding conditions, the corrosive reagent, the immersion time and temperature, and the appearance of the joints before and after immersion.

Sandwich Structures. Test procedures have also been developed to determine the properties of adhesive-bonded structures. However, because of the specialized nature of these tests, they will not be discussed at length here. Most of these procedures are concerned with the evaluation of sandwich structures, particularly those structures used in aircraft construction. Standard test methods are available to determine the following properties of sandwich structures: peel strength, edgewise and flatwise compressive strength, delamination strength of honeycomb core material, flexure and flexure-creep strength, shear fatigue strength, shear strength, and tensile strength. Additional methods are available for measuring core density and core thickness. These methods are discussed in the *ASTM Book of Standards*, Part 16.

REFERENCES

1. R. Houwink and G. Solomon, eds., *Adhesion and Adhesives*, Vol. 2, 2nd Edition (New York: Elsevier Publishing Company, 1967).
2. H. A. Perry, *Adhesive Bonding of Reinforced Plastics* (New York: McGraw-Hill Book Company, Inc., 1959).
3. "Structural Sandwich Constructions; Wood; Adhesives," *Book of ASTM Standards*, Part 16 (Philadelphia: American Society for Testing and Materials, 1968).

chapter 7

case histories

The various case histories in this chapter are included to illustrate some of the advantages of using adhesives in modern manufacturing. Comparisons are made between the various conventional methods of fabrication and the adhesive method to indicate the line of reasoning that was instrumental in making the decision to use adhesives.

STRUCTURAL BONDING OF INNER TO OUTER HOOD AND TRUNK COMPARTMENT PANELS IN THE AUTOMOTIVE INDUSTRY

Beginning with the 1960 model year, the automotive industry began using a heat curing, semi-structural, mastic adhesive for bonding inner and outer panels of hoods and trunk compartment lids. This adhesive is unique in many respects. It is oil tolerant, dispensable from highly automated pumps and ejection applicators, and because it is used in conjunction with spot welding on the same part, it can be applied in a parts stamping plant and heat cured hundreds of miles away in the paint ovens of an automotive assembly plant. In fact, the heat curing operation can take place at any time from a few hours to several weeks later without special handling or fixtures. The adhesive bonding process joins the large outer panel of the hood or trunk compartment lid with the inner reinforcement panel to form a single high strength unit. Although pioneered by only one of the automotive companies on their 1960 hood, today the bonding of hoods and trunk compartment lids is a common assembly practice used by all the U. S. automotive manufacturers.

The primary reason for developing the adhesive and methods for bonding hoods and trunk compartment lids was to eliminate the flutter and rattle that occurred with these components at high speeds or over rough roads. In addition to eliminating the rattle problem, adhesive bonding also offered other advantages. For example, the weight of hood and trunk compartment lid assemblies was reduced; larger and more complex parts requiring high strength reinforcements could be designed. Manufacturing complexity was also reduced by eliminating sprayed-on asphalt sound deadeners and asphalt impregnated anti-rattle pads which were previously hand bonded between the inner and outer panels (1) (2). In addition to these advantages, a significant manufacturing savings of $.14 per unit was realized.

Prior to developing the present vinyl-base mastic adhesive for the automotive industry, numerous types of adhesives were evaluated and screened by automotive materials engineers. Laboratory and proving grounds tests were performed for identifying critical material properties required for assembly processing and service for customers. Because of the careful attention given to the adhesive development and establishing engineering specifications for hood and trunk compartment lid bonding, the following processing and performance requirements are being met with adhesives furnished by major automotive adhesive suppliers:

1) The adhesive is compatible with high volume automatic ejector dispensing equipment and will not cause stoppages because of material deterioration or plugging of ejector gun nozzles.
2) Ingredients in the adhesive have a low order of toxicity for the prevention of dermatitis effects to exposed handlers. Excess material is easily removed with standard cleaning methods and low toxicity solvents.
3) The adhesive does not flow or "slump" from its applied position on the hood or trunk compartment lid during application, storage, shipping, or when the part is subjected to heat exposure for curing.
4) The adhesive will "wet" and adhere to metal surfaces that are contaminated with mill oil and drawing compound from metal stamping operations.
5) After only a short air dry period, the adhesive is not "washed off" or displaced when hoods or trunk compartment lids are subjected to the metal cleaning and zinc phosphate coating process used to prepare sheet metal parts for painting.
6) The adhesive cures at the 275° to 350°F ambient baking temperatures used by the assembly plants.
7) A high level of metal-to-metal bond strength is achieved, and this strength is retained after exposure to a wide range of temperature and humidity environments.
8) Excellent adhesion is also obtained at low temperatures to prevent bond failures when hoods and trunk compartment lids are slammed during winter months.
9) The sound and vibrational damping properties of the adhesive are retained for the life of the vehicle (2).

By using a mastic adhesive, rather than sprayed-on deadeners or hand applied anti-rattle pads, the assembly process was readily automated. The automotive stamping plant can process 400 to 600 hoods or trunk compartment lids per hour, using the mastic adhesive under normal operating conditions. Adhesive is applied to between 80 and 95 spots at predetermined locations on the outer panel, depending on the size of the part. The adhesive application is coordinated with the contact areas on the inner reinforcement panel (hat sections) to insure proper metal to adhesive contact (see Fig. 7–1). The following is a step-wise description of the salient features of the hood and trunk compartment lid assembly process using mastic adhesive (3):

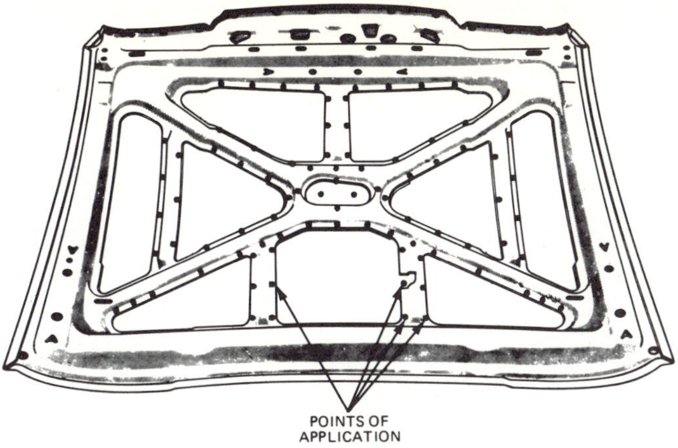

Fig. 7–1. A typical inner and outer panel assembly showing locations at which the structural mastic adhesive is applied. *(Courtesy, Ford Motor Company)*

1) An outer panel from the stamping dies is loaded manually onto a conveyor belt. No attempt is made to remove preservative mill oils or drawing compounds.
2) As the panel moves under the adhesive ejector guns, it is raised into position by a lift table, and the adhesive is automatically applied in the form of numerous conical "spots" (see Fig. 7–2). Adhesive application takes only a few seconds after which the panel continues on the conveyor.

Fig. 7–2. Numerous automatic ejector guns apply 80 to 95 spots of adhesive to the outer panel in a high speed production operation. *(Courtesy, Ford Motor Company)*

3) From an intersecting conveyor line, the inner reinforcement panel is delivered and manually joined to the outer panel (see Fig. 7–3). Each spot of adhesive is compressed by the inner panel insuring good contact and "wetting" of the bonding surfaces.

Fig. 7–3. After adhesive application, outer and inner reinforcement panels are combined in preparation for final spot welding. *(Courtesy, Ford Motor Company)*

4) After joining, the assembly continues down the conveyor through clinch dies where the hem (edge) flanges of the outer panel are turned and flattened against the inner panel.
5) The final assembly operation consists of spot welding the inner and outer panel together around the edges of the hem flanges, integrally joining the components into a single unit. Spot welding is necessary for assembly and serves two purposes; first, it aids the final product structurally; secondly, it allows the assembly itself to become a fixture to hold the inner and outer panels in place until the hood or trunk compartment lid is processed through the assembly plant paint oven to cure the adhesive (see Fig. 7–4).

Since the first hood was bonded in 1960, designers and materials engineers throughout the industry have developed many uses for the mastic adhesive, such as bonding interior reinforcements to door panels, and bonding framing members to the inside of van type vehicles. Bonding with mastic adhesive is regularly considered where added strength is desired in a metal-to-metal assembly without increasing the metal gauge, or where other methods of assembly, such as welding, would create metal finishing problems (4).

The adoption of the mastic adhesive by the automotive industry for bonding hoods and trunk compartment lids banished many myths and prejudices against the use of adhesive for mass production bonding operations. This success illustrates very well that when proper part design, engineering, processing and adhesive are combined, a better, lower cost assembly can result.

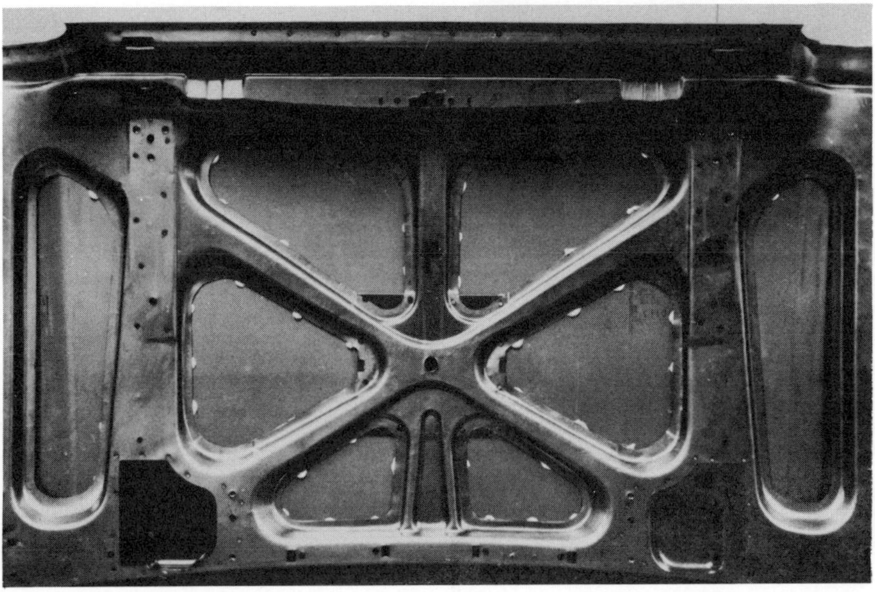

Fig. 7–4. Completed assembly showing the adhesive squeezed out between outer and inner panel indicating good bonding contact. *(Courtesy, Ford Motor Company)*

BONDING IN THE BOAT BUILDING INDUSTRY

Characteristics of the Adhesive

Certain applications for sealants in the small boat industry require that these sealants perform as adhesives and that they withstand intermittent contact with fresh and salt water. The performance of polysulfide liquid polymer-based sealants has resulted in considerably reducing the average annual maintenance expense which is an important factor for the small boat owner.

In the manufacture of 26 foot sea skiffs using lapstrake construction on the hull (see Fig. 7–5), the oldest manufacturing facility uses a polysulfide liquid polymer-based sealant. This first application in 1953 has remained trouble free for 15 years with no leakage and has never been recaulked or a single lap refastened. Such performance is extraordinary considering the problems that occur as a result of placing boats in water after a winter in dry dock. This annual cycle of dry winters and wet summers causes considerable movement in most joints on small boats and normally requires annual or biennial recaulking of hull joints using conventional caulking.

The use of polysulfide base sealants to fill construction joints has eliminated the need for swelling the wood in order to close joints. Keeping joints dry has

Fig. 7–5. Completed sea skiff sealed in 1953 using full lapstrake construction. *(Courtesy, Thiokol Chemical Corporation)*

also minimized or eliminated dry rot which can occur with conventional caulking practices.

The sealant-adhesive is formulated from polysulfide liquid polymer, selected fillers, and additives to give a semi-fluid compound which converts to a rubber either by the use of a catalyst or by the action of moisture from the air. The converted compound shrinks less than two percent during cure and remains as a rubber irrespective of time, temperature, or water immersion. Heat aging of cured sealant even after one week at 212°F results in less than five percent weight loss or more than a 20 point increase in rubber hardness. Adhesion assemblies made from various woods show no loss of adhesion even after long periods of continual immersion in fresh or salt water. The sealant remains flexible at temperatures down to −40°F and is unaffected by gasoline, oil, grease, or various automotive fluids that might be present in the bilge area.

The sealant-adhesive is formulated to give a flexible rubbery compound which will withstand movement and yet behave as an adhesive. The general practice in boating is to use polysulfide sealant in combination with mechanical fasteners.

The use of polysulfide sealants is not limited to lapstrake construction, and many other applications have been found which have greatly reduced the average annual maintenance normally required for small boats. These are listed as follows:

1) Sealing stem and keel joints
2) Sealing hatch and skylight frames
3) Sealing hull seams
4) Sealing to transom joints
5) Glazing windows and port lights
6) Caulking deck seams
7) Bedding deck hardware
8) Bedding canvas deck covers
9) Bedding chines and sheer clamps
10) Bedding keelson, bedlog, and bottom planking on the centerboard well.

Method of Assembly

Polysulfide liquid polymer based sealants are available as a two-part catalyzed sealant or a one-part sealant that cures upon contact with moisture from the air. The two-part sealants are used when a fast cure is required, and the sealant may be embedded between components. Here, the sealant is formulated so that after mixing with the catalyst, the sealant has at least four hours of work life. This sealant will be cured in two days irrespective to its location and its mass. The one-part sealants cure upon exposure to the atmosphere and will become tack-free in one day. However, one-part sealants may cure through in one to four weeks in cross-sections ½ in. × ½ in. and consequently are best used on above deck area, or for sealing joints when the boat is in dry dock, and time is not a factor.

When a two-part sealant is used in the plant, the two-part sealant may be automatically metered, mixed, and applied on sections of lapstrake siding prior to attachment. For isolated installations, the mixed two-part sealant can be pumped into cartridges for use in handguns for the various miscellaneous installations.

If the sealant is to be used between layers of siding, the priming of wood may not be required. If the sealant is to be used in joints between deck planks (see Fig. 7–6) or in the hull (see Fig. 7–7), priming of wood is recommended for continual adhesion in spite of water immersion.

Fig. 7–6. Application of polysulfide sealant on deck joints. (*Courtesy, Thiokol Chemical Corporation*)

Cost Considerations

An average 26 foot sea skiff requires sealant in the application areas as shown in Table VII–1 (see also Figs. 7–8 and 7–9).

The cost calculations in Table VII–1 do not include labor, since labor is assumed to be identical, irrespective of the type of caulking compound used. If recaulking is repeated periodically, the cost of labor becomes a major expense. Generally, recaulking must be carried out every two years using a conventional caulk as compared to the use of polysulfide sealant in the original installation, which might have to be replaced after 16 years. On this basis (frequency of repair and maintenance), Table VII–2 compares costs using conventional and polysulfide sealant systems.

Fig. 7–7. Application of polysulfide sealant along the overlapping edge of lapstrake board. (*Courtesy, Thiokol Chemical Corporation*)

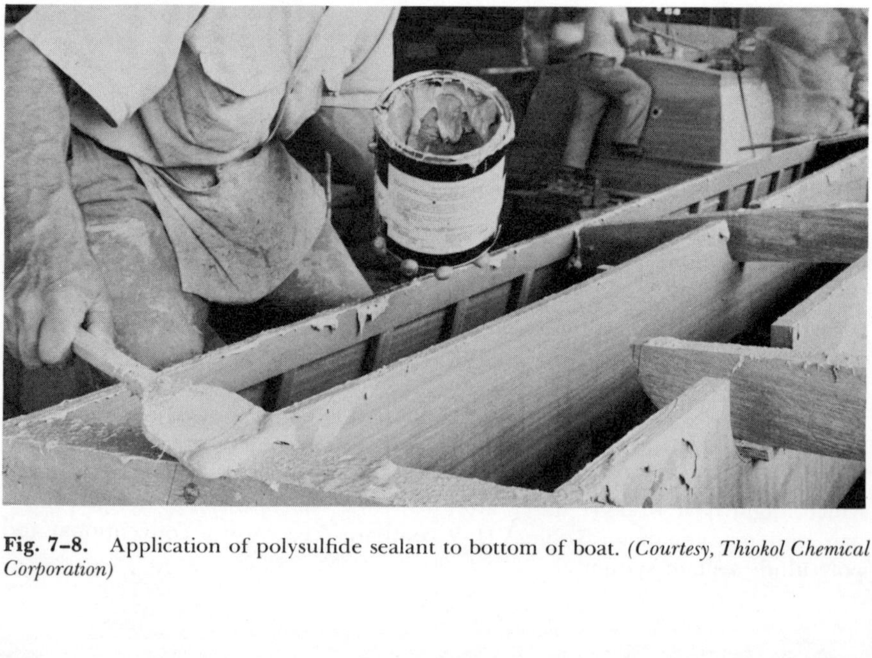

Fig. 7–8. Application of polysulfide sealant to bottom of boat. (*Courtesy, Thiokol Chemical Corporation*)

Fig. 7–9. Application of polysulfide sealant to joints along the beam. *(Courtesy, Thiokol Chemical Corporation)*

Table VII–1. Typical Sealant Requirements on an Average 26 foot Sea Skiff.

Application Area	*Total Requirement*	*Sealant Application Rate*	*Total Gallons*
Lapstrake sealing	1200 linear ft	75 ft/gal	16.0
Bedding canvas covers	300 ft^2	100 ft^2/gal	3.0
Caulking deck seams	300 linear ft	300 ft/gal	1.0
Caulking windows	200 linear ft	400 ft/gal	0.5
Bedding clamps, chines, etc.			2.0
			22.5 gal

Average price of polysulfide sealant \$15/gal × 22.5 gal	\$337.50
Average price of marine sealant \$5/gal × 22.5 gal	112.50
Price increase due to polysulfide sealant	\$225.00

Table VII–2. Comparison of Costs Using Temporary vs Permanent Sealant in Sealing a Lapstrake Hull.

Conventional Sealant – replaceable every 2 years	
16 gal sealant – 8 installations at \$5/gal	\$ 640
Labor for recaulking 1,200 linear feet – 8 installations at \$.25/linear ft	2,400
Total cost	\$3,040
Polysulfide Sealant – replaceable every 16 years or more	
16 gal sealant – 2 installations at \$15/gal	\$ 480
Labor for recaulking 1,200 linear feet – 2 installations at \$.25/linear ft	600
Total cost	\$1,080
Savings \$1,960	

A comparison of the initial costs of conventional sealant vs polysulfide sealant in Table VII–2 runs $380 for conventional and $540 for polysulfide. This includes $300 for cost of labor which is the same for the application of each sealant. But the low cost conventional caulked boats will have to be probably recaulked every two years in many of the joints. Also, the above comparison does not cover possible additional costs of maintenance and repair in other areas, so that Table VII–2 would be considered minimal.

In the original manufacture of small boats, the use of polysulfide sealant in all areas of caulking results in an essentially maintenance free product for a number of years. Even when some repair might be needed, the sealant does not have to be replaced, and only occasional joints need be recaulked where unusual stress or movement took place.

Safety Requirements and Regulations

Polysulfide liquid polymer has been tested for oral and skin toxicity and was found to be nonhazardous according to the interpretation of the Federal Hazardous Substances Labeling Act. Moreover, an average polysulfide-based sealant has also been tested and found nonhazardous.

Most sealants contain little or no solvent which might cause skin irritation. The acute oral toxicity is above the minimum tolerance of the Labeling Act. Eye irritation does not show any unusual involvement. Consequently, only normal clean household practices are needed when working with average polysulfide sealants, and although not necessary, cloth gloves are recommended to reduce cleanup.

Although highly improbable, if polysulfide sealant should be accidentally ingested, it is suggested that an emetic be used to remove the contaminant from the body.

Unused sealant can be dissolved in toluol or xylol, but cured sealant must be physically cut from any undesirable area since it is not affected by solvent.

FOOD PACKAGING: HOT MELT ADHESIVES

Hot-melt adhesives are being used increasingly in food packaging for everything from carton, can, pouch, and bag sealing to the fabrication of spiral-wound containers and the lamination of films, papers, and foils.

Characteristics

The reason for the success of hot melt adhesives is the need for adhesives that "set up" almost instantly on ultra-fast packaging or fabricating machines and that form bonds impervious to moisture, heat, or product constituents. All of these conditions can be met by these solid adhesives, which melt for application then revert to their solid state upon cooling to complete the adhesive bond. Specifically, the significant attributes of hot melt adhesives which make them useful in the high volume food packaging lines are as follows:

1) Greater use with hot melt adhesives is frequently the case, depending on the specific application, when compared with water-based compounded dextrin and polyvinyl acetate adhesives.
2) Production speed is increased with hot melts over conventional solution

or emulsion systems. For the latter, the minimum setting or bonding speed is four to five seconds while hot melts can set in less than one second.

3) Reduction in the floor space requirements in packaging or fabricating lines because of the fast set of the hot melts. When hot melts are used in a typical case closing operation, a six ft compression section replaces a 35 foot wet adhesive unit.
4) Lower labor and maintenance costs result when hot melts replace the standard wet adhesives. In the case closing example, down time and maintenance costs of a wet adhesive system totaled 26 man hr/wk for cleanup of the equipment compared to less than 5 man hr/wk when hot melts were used.
5) Adhesive waste and bond failure due to humidity and solvent variation are eliminated when hot melts are used in place of wet adhesives.
6) Hot melts are versatile, i.e., they can be used to bond impervious substrates, requiring only cooling to make the bond. Emulsions are unsuitable because the solvent diluents cannot escape into the impervious substrates to set the bond.

Safety Considerations

No special precautions outside of avoiding contact with the hot adhesive or associated equipment are necessary when using hot melts.

FDA Approval

In food packaging, hot melt adhesives must meet the requirements described in the Federal Register in paragraph 121.2520, Subpart F, Chapter 1, Title 21 of the Food, Drug and Cosmetic Act. Under this Act, these adhesives will be labeled as Food Packaging Adhesives and can be safely used for that purpose.

Case Histories

Under food packaging applications, three case histories have been selected to illustrate some of the advantages which result when hot melt adhesives are used in place of conventional wet adhesives or mechanical closures.

Case A: Non-Destructive Carton Opening. Carton sealing of cereal box tops was normally made with wet adhesives. For years this was acceptable; however, when the boxes were modernized to have a reclosure feature, easy opening became a necessity. No longer was it permissible to bond the box top shut and then tear it open without regard to flap integrity since this would destroy the reclosure feature of the new box tops. Attempts to solve this problem using wet glues by printing a pattern of releasing ink to the box top to enable some control of bonding, i.e., the wet adhesive would bond only to the unprinted surface, was not entirely successful. Finally, the need for faster production guided the decision to use hot melt adhesives. The choice was made because they give: a consistent easy opening box top without destruction of the box top, easy control of the force required to open the box top by controlling the size of the dots of the hot melt adhesive, fast set (cutting the compression section necessary for setting of the wet adhesives from 20 ft to 2 ft), and a reduction in the handling costs when buying, shipping, and storing 100 percent solid adhesive.

The 40 cent cost of a pound of adhesive for sealing 10,000 box tops was even less than the cost for the water-based adhesive and was also instrumental in making the decision.

Case B: Simplified Less Expensive Package. A frozen food packager needed a high speed packaging system. Three applicators designed to accommodate continuous high-speed packaging operations were installed to handle a hot melt adhesive in cord-like form from a reel. The adhesive was melted electrically as required. The conversion to the new adhesive system resulted in the following advantages:

1) Efficiency. Elimination of the pressure-drying cycle and the long compression section needed for wet adhesive; elimination of glue pots and resulting daily clean-up.
2) Cost. All logo, product illustrations, and cooking instructions were on the carton eliminating the need for wax or foil overwrap as was used with wet adhesives to carry this printed material.
3) Consumer convenience. Without the overwrap and with controlled sealing of carton flaps, the carton could be easily and quickly opened without damaging the cooking instructions which would normally occur when the overwrap packages were torn open.

Case C: Overpackaged Product. This example illustrates a cost saving idea on packaging a product which could produce an annual savings of over \$50,000 on production of over a million packages of frozen potato products a week.

Originally, the box top for the package was mechanically closed. It required a heavier box than necessary to hold the food and prevent the box from collapsing during the closing operation.

A 20 percent cost savings could be obtained if cardboard 2 mils thinner were used and if the tops were glued instead of being mechanically closed.

Of all the adhesives considered, only the hot melts would work fast enough. The others would require a machine with a long compression section, unavailable because of a lack of floor space for extension of the assembly lines. The changeover to a hot melt bonding system was made and was so successful that the anticipated saving of \$50,000 for the year was realized. The cost of the new applicators were written off in the first 10 weeks after conversion.

The hot melt units easily met the following requirements:

1) Easily adaptable to existing closure sections. Each unit was installed in less than one day requiring a minimum of downtime and on only one line at a time. The compression section for the hot melt system was only $1\frac{1}{4}$ ft whereas 30 ft would be needed if a water-based adhesive were used.
2) Easy removal of the hot melt units should it be necessary to revert to the conventional top-closure system.
3) Provide a high degree of reliability for continuous or intermittent use in a plant operating 3 shifts, 6 days a week, year round. The adhesive applicators held the adhesives at a constant temperature and recycled the adhesive by-passing the extruder head should it be necessary to stop the line for any reason. Thus the equipment does not cool down between the 3 shifts, and requires no warm-up time.
4) Accurate lay down of an inch-long bead of adhesive on one of the top flaps of each carton with no skipping or stringing. Adhesive usage was

minute; only about one pound was needed for every 30,000 cartons and material cost was only 1½ cents for 1,000 cartons.

5) Capability of almost flawless performance at high speed (150–180 cartons/minute) and even higher production rates if necessary in the future.

POLYSULFIDE GLASS BONDING FOR AUTOMOTIVE USE

The following is a case history involving the bonding of back windows (see Fig. 7–10) to metal fence openings in an automotive assembly plant. Prior to the use of the multicomponent polysulfide system, a rubber weatherstrip was used.

Fig. 7–10. Installation of the back window on an automotive assembly line. *(Courtesy, Chemical Division, Chrysler Corporation)*

The weatherstrip was designed with two channels, one for placement into the metal fence and the other to receive the glass edge of the back window. In addition, a locking strip was provided in the rubber design to close the openings and to provide a finished appearance.

The utilization of the polysulfide produces a bonded glass to metal assembly with the following noted advantages:

1) A substantial cost savings was realized by the reduction of manpower previously required for the numerous sealing operations in the rubber weatherstrip
2) A significant improvement in appearance was obtained by the flush fitting of exterior trim moldings to the glass. (The rubber spacing between the moldings and glass have now been eliminated)
3) A reduction in water leaks has substantially reduced customer complaints in this area and has also lowered warranty costs.

The polysulfide sealant used for the bonding of glass consists of three parts: the base material and the fast and the slow accelerators, each of which are supplied in separate containers. The base material, received in 50 gal drums, is heavy in consistency; the accelerators are received in 5 gal pails. However, the thixotropic properties of the base material allow the sealant to thin out under pumping pressures thereby permitting an adequate application of material. The removal of pressure returns the polysulfide sealant to its original heavy consistency. Material flow is then prevented which would seriously impair the application of this product.

The accelerators, also called curing agents, consist of lead peroxide* and are formulated to regulate the rate of cure when blended with the base material. The total amount of base to accelerator is 10:1 parts by weight. The addition of only the fast accelerator, a combination of both, or the addition of just the slow accelerator to the base product produces a usable work life period of from 10 min to 3 hours depending upon each plant's requirement. This is determined by the time period between adhesive application and various operations such as trim installation, cleaning operations, door slamming, and water tests, which should be conducted only after adequate curing.

In the application of the material, all three components are extruded through metering mixing units (see Fig. 3–7). When the proper accelerator ratios have been determined, the mixing head of the equipment meters the two curing agents into the base material through a common applicator gun (see Fig. 7–11). Cold water cooling is utilized to prevent a temperature buildup in

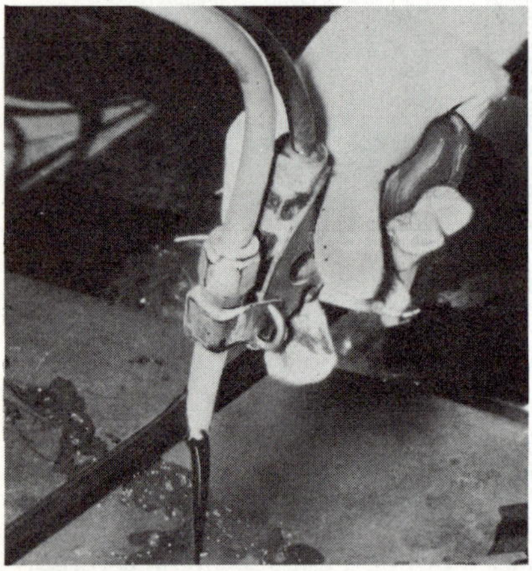

Fig. 7–11. Applicator gun. *(Courtesy, Chemical Division, Chrysler Corporation)*

the mixing head. In addition, purging of equipment including the gun and hoses used in this application should be strictly maintained to eliminate curing of the material between production shifts and during downtimes.

Plant Precautions Required in the Use of the Polysulfide

Humidity and temperature fluctuations can cause pumping problems, material flow, and changing cure characteristics. In general a rise in temperature of 10°F will cut the work lifetime by 10 min. A similar rise in humidity will have the

*A toxic chemical. Safety considerations should be carefully studied. Check safety instructions from the manufacturer. See also Chapter 3.

same effect but to a lesser degree. Temperature conditions are usually maintained at 70°F plus or minus 2°F.

The shelf life of the base material being three months does not create a problem since proper rotation of stock will adjust any excess inventory. The accelerators, however, require agitation prior to placement under the pumping system due to their settling characteristics.

Contamination of the primer with dirt or other impurities can cause poor bonds either to the glass or to the metal surface. Therefore, it is important that all containers of this material be properly sealed when not in use.

The height of the polysulfide bead as well as the amount should be controlled when it is extruded onto the glass; also, complete contact should be made with the metal fence opening.

Regulated purging and periodic cleaning of the equipment is required to prevent material cure in the mixing head resulting in production line stoppages.

The equipment area should be screened off and locked when not in use. Tampering with the equipment and its controls by other than qualified personnel could result in variations which could affect the physical cure properties of the material.

Smoking or the carrying of lighted matches should not be allowed in the confines of the equipment area because of the volatile solvents necessary for cleaning. Also, skin contact is to be avoided in the handling of the accelerators.

Adhesive bonding can be as simple as attaching various insulating or trim materials to metal with spray adhesives. On the other hand, the application might be as involved as bonding glass to a finished assembly. In the latter case, the more costly mechanical attachment of glass to metal has been eliminated while at the same time the polysulfide application has substantially improved the car's appearance and has, at the same time, provided its own sealing.

In spite of the various provisions which must be adopted to successfully utilize the polysulfide in production, a highly successful and profitable operation has been realized.

AIRCRAFT—BONDING OF SANDWICH PANELS

The bonding of sandwich panels in aircraft has become more common over the years. This case history discusses the bonding of sandwich panels for the leading edge slats on a heavy transport aircraft (see Fig. 7–12). The case involves a comparison between (1) a conventional leading edge with a honeycomb trailing edge wedge, and (2) a complete honeycomb panelized structure with sheet metal support ribs.

Due to the difficulties experienced with contour machining and bonding of full-depth contoured assemblies having core depths above 6 in., the leading edge slats were originally designed with a conventional leading edge combined with a honeycomb trailing edge wedge of a maximum depth of 4½ in. The slat leading edge which is a sheet metal skin supported by a beam and rib structure is joined to the honeycomb trailing edge by mechanical fasteners (see Fig. 7–13). The honeycomb close-out members and the joint of the trailing edge to leading edge were, on review, considered to be unnecessary structural weight. Several alternate designs were studied considering minimum weight structure and optimum producibility. The final selection of the design is shown in Fig. 7–14. This

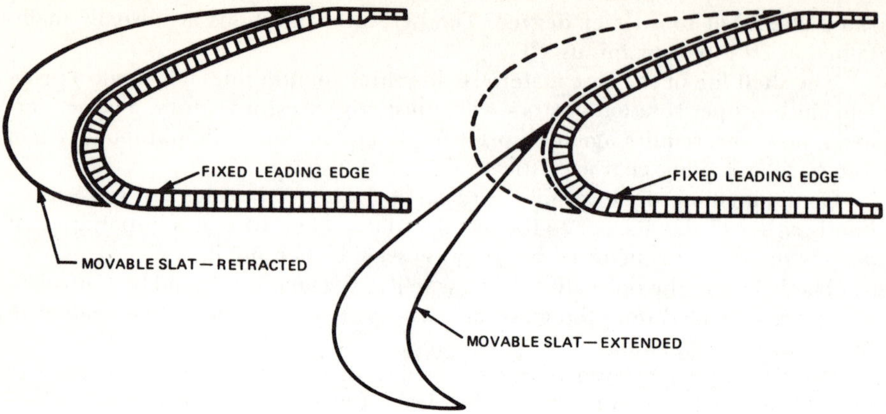

Fig. 7–12. Sketch of fixed leading edge and moving slat geometry.

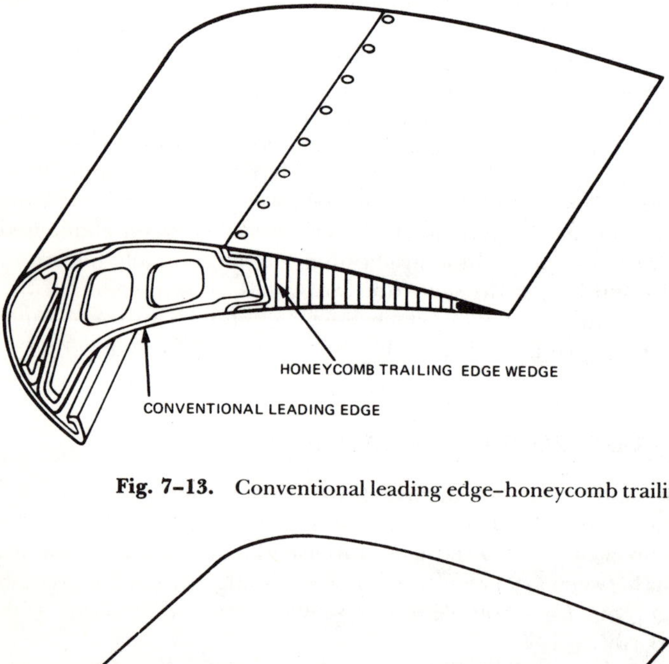

Fig. 7–13. Conventional leading edge–honeycomb trailing edge wedge.

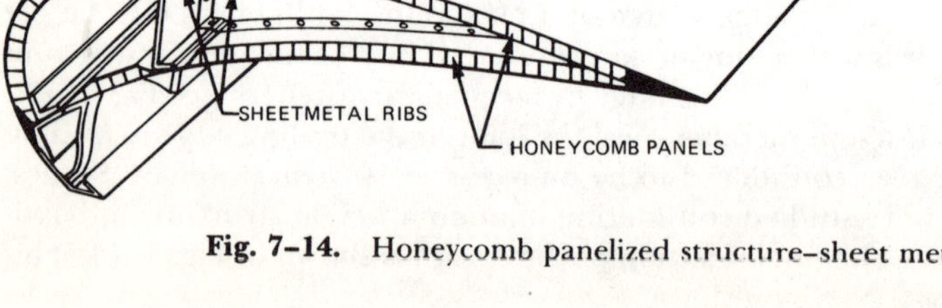

Fig. 7–14. Honeycomb panelized structure–sheet metal ribs.

design combined two contoured honeycomb panels that were joined at their lower leading edge and at the trailing edge and supported by a beam and rib sheet metal structure.

Comparison of the two methods of fabrication reveals that the honeycomb panelized structure eliminates the trailing edge beam and the leading edge to the trailing edge joint. The structural strength of the honeycomb panels helps to cut the weight of the supporting structure and the gage of the skin.

The conventional design had a calculated weight of 1064 lbs for three slats whereas the honeycomb panelized structure has a calculated weight of 874 lbs for the same three slats of the assembly. The final bonding technique saved a total weight of 190 lbs for the three slats.

REFERENCES

1. "Adhesive-Bonding Hood Assemblies," *American Machinist/Metalworking Manufacturing,* August, 1962.
2. D. L. Jones and E. P. DePalma, "A Tailor-Made High Product Metal to Metal Automotive Adhesive," *SAE Technical Paper,* March, 1960.
3. E. P. DePalma, "Topic No. 25," (Paper for ASM Metals Handbook Committee on Adhesive Bonding of Steel, American Society for Metals, September 14, 1964).
4. Milton Alexander, ed., "The Case for Metal to Metal Adhesive Bonding," *Production,* Vol. 57, No. 5 (May, 1966), pp. 112–23.

glossary

A-Stage—An early stage in the reaction of a thermosetting resin in which the material is still soluble in certain liquids and fusible. (See also *B-Stage* and *C-Stage.*)

Adhere—To cause two surfaces to be held together by adhesion.

Adherend—A body which is held to another body by an adhesive. (See also *Substrate.*)

Adhesion—The state in which two surfaces are held together by interfacial forces which may consist of valence forces or interlocking action, or both. (See also *Adhesion, Mechanical* and *Adhesion, Specific.*)

Adhesion, Mechanical—Adhesion between surfaces in which the adhesive holds the parts together by interlocking action. (See also *Adhesion, Specific.*)

Adhesion, Specific—Adhesion between surfaces which are held together by valence forces of the same type as those which give rise to cohesion. (See also *Adhesion, Mechanical.*)

Adhesive—A substance capable of holding materials together by surface attachment.

Adhesive, Assembly—An adhesive which can be used for bonding parts together, such as in the manufacture of a boat, airplane, furniture, and the like.

Adhesive, Cold-Setting—An adhesive which sets at temperatures below 68°F (20°C). (See also *Adhesive, Hot-Setting; Adhesive, Intermediate Temperature Setting;* and *Adhesive, Room Temperature Setting.*)

Adhesive Dispersion—A two-phase system in which one phase is suspended in a liquid.

Adhesive, Hot-Setting—An adhesive which requires a temperature at or above 212°F (100°C) to set it. (See also *Adhesive, Cold-Setting; Adhesive, Intermediate Temperature Setting;* and *Adhesive, Room Temperature Setting.*)

Adhesive, Intermediate Temperature Setting—An adhesive which sets in the temperature range of 87° to 211°F (31° to 99°C). (See also *Adhesive, Cold-Setting; Adhesive, Hot-Setting;* and *Adhesive, Room Temperature Setting.*)

Adhesive, Pressure-Sensitive—An adhesive that adheres to a surface at room temperature by briefly applied pressure alone.

Adhesive, Room Temperature Setting—An adhesive which sets in the temperature range of 68° to 86°F (20° to 30°C), in accordance with the limits for Standard Room Temperature specified in the Standard Methods of Conditioning Plastics and Electrical Insulating Materials for Testing (ASTM Designation: D-618). (See also *Adhesive, Cold-Setting; Adhesive, Hot-Setting;* and *Adhesive, Intermediate Temperature Setting.*)

Adhesive, Separate Application—A term used to describe an adhesive consisting of two parts, one part being applied to one adherend and the other part to the other adherend, and the two brought together to form a joint.

Adhesive, Warm-Setting—A term which is sometimes used as a synonym for *Intermediate Temperature Setting Adhesive.* (See *Adhesive, Intermediate Temperature Setting.*)

Aging Time—See *Time, Joint Conditioning.*

Aggressive Tack—See *Tack, Dry.*

Alloy (polymer)—An intimate blend or a copolymer of two individual polymer types.

Amylaceous—Pertaining to, or of the nature of, starch; starchy.

Anchor Pattern—A pattern made by blast cleaning abrasives on an adherend surface in preparation for adhesive application prior to bonding. Pattern is examined in profile.

Assembly—A group of materials or parts, including the adhesive, which has been placed together for bonding or which has been bonded together.

Assembly Adhesive—See *Adhesive, Assembly.*

Assembly Glue—See *Adhesive, Assembly.*

Assembly Time—See *Time, Assembly.*

B-Stage—An intermediate stage in the reaction of a thermosetting resin in which the material softens when heated and swells in contact with certain liquids, but may not entirely fuse or dissolve. Resins in thermosetting molding compounds are usually in this stage. (See also *A-Stage* and *C-Stage.*)

Bag Molding—A method of molding or bonding involving the application of fluid pressure, usually by means of air, steam, water, or vacuum, to a flexible cover which, sometimes in conjunction with the rigid die, completely encloses the material to be bonded.

Binder—A component of an adhesive composition which is primarily responsible for the adhesive forces which hold two bodies together. (See also *Extender* and *Filler.*)

Blister—An elevation of the surface of an adherend, whose shape somewhat resembles a blister on the human skin; its boundaries may be indefinitely outlined, and it may have burst and become flattened.

Blocking—An undesired adhesion between touching layers of a material, such as occurs under moderate pressure during storage or use.

Bond—The attachment at an interface between an adhesive and an adherend. (See also *Joint.*) To attach materials together by adhesives.

Bond Strength—The unit load applied in tension, compression, flexure, peel, impact, cleavage, or shear, required to break an adhesive assembly with failure occurring in or near the plane of the bond.

C-Stage—The final stage in the reaction of a thermosetting resin in which the material is relatively insoluble and infusible. Thermosetting resins in a fully cured adhesive layer are in this stage. (See also *A-Stage* and *B-Stage.*)

Catalyst—A substance which markedly speeds up the cure of an adhesive when added in minor quantity as compared to the amounts of the primary reactants. (See also *Hardener* and *Inhibitor.*)

Cement—See *Adhesive.*

Cement—See *Bond.*

Cps—Centipoise, a unit of viscosity equal to $^1/_{100}$ poise.

Closed Assembly Time—See *Time, Assembly.*

Cohesion—The state in which the particles of a single substance are held together by primary or secondary valence forces. As used in the adhesive field, the state in which the particles of the adhesive (or the adherend) are held together.

Cold Flow—See *Creep.*

Cold Pressing—A bonding operation in which an assembly is subjected to pressure without the application of heat.

Cold-Setting Adhesive—See *Adhesive, Cold-Setting.*

Colophony—See *Rosin.*

Condensation—A chemical reaction in which two or more molecules combine with the separation of water or some other simple substance. If a polymer is formed, the process is called polycondensation. (See also *Polymerization.*)

Conditioning Time—See *Time, Joint Conditioning.*

Consistency—That property of a liquid adhesive by virtue of which it tends to resist deformation.

Copolymer—See *Polymer.*

Copolymerization—See *Polymerization.*

Crazing—Fine cracks which may extend in a network on or under the surface of or through a layer of adhesive.

Creep—The dimensional change with time of a material under load, following the initial instantaneous elastic or rapid deformation. Creep at room temperature is sometimes called *Cold Flow.*

Cross Laminated—See *Laminated, Cross.*

Crosslinking—The process of chemically connecting individual polymer chains at several points to form a three-dimensional network.

Cryogenic—Relating to the effects of very low temperatures.

Cure—To change the physical properties of an adhesive by chemical reaction, which may be condensation, polymerization, or vulcanization; usually accomplished by the action of heat and catalyst, alone or in combination with or without pressure.

Curing Time—See *Time, Curing.*

Delamination—The separation of layers in a laminate because of failure of the adhesive, either in the adhesive itself or at the interface between the adhesive and the adherend, or because of cohesive failure of the adherend.

Diluent—An ingredient, usually added to an adhesive to reduce the concentration of bonding materials. (See also *Extender.*)

Doctor-Roll—A roller mechanism which revolves at a different surface speed, or in an opposite direction, resulting in a wiping action for regulating the adhesive supplied to the spreader roll.

Doctor-Bar or *Blade*—A scraper mechanism which regulates the amount of adhesive on the spreader rolls or on the surface being coated.

Double Spread—See *Spread.*

Dry—To change the physical state of an adhesive on an adherend by the loss of solvent constituents by evaporation or absorption, or both. (See also *Cure* and *Set.*)

Dry Strength—See *Strength, Dry.*

Dry Tack—See *Tack, Dry.*

Drying Time—See *Time, Drying.*

Elastomer—A material which at room temperature can be stretched repeatedly to at least twice its original length and, upon immediate release of the stress, will return with force to its approximate original length.

Extender—A substance, generally having some adhesive action, added to an adhesive to reduce the amount of the primary binder required per unit area. (See also *Binder* and *Filler.*)

Faying Surface—An object's surface coming in contact with another surface to which it is bonded.

Filler—A relatively nonadhesive substance added to an adhesive to improve its working properties, permanence, strength, or other qualities. (See also *Binder* and *Extender.*)

Flow—Movement of an adhesive during the bonding process, before the adhesive is set.

Functionality—The average number of reaction sites on an individual polymer chain.

Gel—A semisolid system consisting of a network of solid aggregates in which liquid is held.

Gelation—Formation of a gel.

Glue—Originally, a hard gelatin obtained from hides, tendons, cartilage, bones, etc., of animals. Also, an adhesive prepared from this substance by heating with water. Through general use the term is now synonymous with the term "adhesive." (See also *Adhesive, Bond, Mucilage, Paste,* and *Sizing.*)

Gum—Any of a class of colloidal substances, exuded by or prepared from plants, sticky when moist, composed of complex carbohydrates and organic acids, which are soluble or swell in water. (See also *Adhesive, Glue, and Resin.*)

Hardener—A substance or mixture of substances added to an adhesive to promote or control the curing reaction by taking part in it. The term is also used to designate a substance added to control the degree of hardness of the cured film. (See also *Catalyst.*)

Hot-Setting Adhesive—See *Adhesive, Hot-Setting.*

Inhibitor—A substance which slows down chemical reaction. Inhibitors are sometimes used in certain types of adhesives to prolong storage or working life.

Intermediate Temperature Setting Adhesive—See *Adhesive, Intermediate Temperature Setting.*

Joint—The location at which two adherends are held together with a layer of adhesive. (See also *Bond.*)

Joint, Lap—A joint made by placing one adherend partly over another and bonding the overlapped portions together. (See also *Joint, Scarf.*)

Joint, Scarf—A joint made by cutting away similar angular segments of two adherends and bonding the adherends with the cut areas fitted together. (See also *Joint, Lap.*)

Joint, Starved—A joint which has an insufficient amount of adhesive to produce a satisfactory bond.

Laminate—A product made by bonding together two or more layers of material or materials to unite layers of material with adhesive. (See also *Laminated, Cross* and *Laminated, Parallel.*)

Laminated, Cross—A laminate in which some of the layers of material are oriented at right angles to the remaining layers with respect to the grain or strongest direction in tension. (See also *Laminated, Parallel.*)

Laminated, Parallel—A laminate in which all the layers of material are oriented approximately parallel with respect to the grain or strongest direction in tension. (See also *Laminated, Cross.*)

Lamination—The process of preparing a laminate. Also, any layer in a laminate.

Lap Joint—See *Joint, Lap.*

Latex—A milk-like fluid in which small globules of natural or synthetic rubber or plastic are suspended in water.

Mechanical Adhesion—See *Adhesion.*

Melt Index—Fluidity of a material as measured by rate of flow through an orifice at a controlled elevated temperature and pressure.

Modifier—Any chemically inert ingredient added to an adhesive formulation that changes its properties. (See *Filler, Plasticizer,* and *Extender.*)

Monomer—A relatively simple compound which can react to form a polymer. (See also *Polymer.*)

Mucilage—An adhesive prepared from a gum and water. Also in a more general sense, a liquid adhesive which has a low order of bonding strength. (See also *Adhesive, Glue, Paste,* and *Sizing.*)

Novolac—A phenolic-aldehydic resin which, unless a source of methylene groups is added, remains permanently thermoplastic. (See also *Resinoid* and *Thermoplastic.*)

Open Time—This is the time interval between the coating of the adhesive on the adherends and the assembly of the components.

Organosol—A colloidal (finely divided) dispersion of a synthetic resin in plasticizer, the volatile content exceeding 5 percent by weight.

Parallel Laminated—See *Laminated, Parallel.*

Paste— An adhesive composition having a characteristic plastic-type consistency, that is, a high order of yield value, such as that of a paste prepared by heating a mixture of starch and water and subsequently cooling the hydrolyzed product. (See also *Adhesive, Glue, Mucilage,* and *Sizing.*)

Penetration—The entering of an adhesive into an adherend.

Permanence—The resistance of an adhesive bond to deteriorating influences.

Pickup Roll— A spreading device immersed and rotating in the reservoir of adhesive to pick up the adhesive.

Plasticity—A property of adhesives which allows the material to be deformed continuously and permanently without rupture upon the application of a force that exceeds the yield value of the material.

Plasticizer—A material incorporated in an adhesive to increase its flexibility, workability, or distensibility. The addition of the plasticizer may cause a reduction in melt viscosity, lower the temperature of the second-order transition, or lower the elastic modulus of the solidified adhesive.

Plastisol—A liquid dispersion of finely divided resin in a plasticizer. It usually has no volatiles. When volatile content exceeds 5 percent, it is called an organosol.

Plywood—A cross-bonded assembly made of layers of veneer or veneer in combination with a lumber core or plies joined with an adhesive. Two types of plywood are recognized, namely: (1) veneer plywood, and (2) lumber core plywood.

Polycondensation—See *Condensation.*

Polymer—A compound formed by the reaction of simple molecules having functional groups which permit their combination to proceed to high molecular weights under suitable conditions. Polymers may be formed by polymerization (addition polymer) or polycondensation (condensation polymer). When two or more monomers are involved, the product is called a copolymer.

Polymerization—A chemical reaction in which the molecules of a monomer are linked together to form large molecules whose molecular weight is a multiple of that of the original substance. When two or more monomers are involved, the process is called copolymerization or heteropolymerization. (See also *Condensation.*)

Pot Life—See *Working Life.*

Prepolymer—An intermediate stage in the preparation of a polymer in which the product is partially formed but is still liquid or is softenable by heating. Analogous to *B-Stage.*

Pressure-Sensitive Adhesive—See *Adhesive, Pressure-Sensitive.*

Primer—A coating applied to a surface, prior to the application of an adhesive, to improve the performance of the bond.

Resin—Any of a class of solid or semi-solid organic products of natural or synthetic origin, generally of high molecular weight with no definite melting point. (See also *Gum.*)

Resinoid—Any of the class of thermosetting synthetic resins, either in their initial temporarily fusible state or in their final infusible state. (See also *Novolac* and *Thermosetting.*)

Resite—An alternate term for *C-Stage* with reference to phenolformaldehyde resins.

Resitol—An alternate term for *B-Stage* with reference to phenolformaldehyde resins.

Resol—An alternate term for *A-Stage* with reference to phenolformaldehyde resins.

Retrogradation—A change of starch pastes from low to high consistency on aging.

Room Temperature Setting Adhesive—See *Adhesive, Room Temperature Setting.*

Rosin—A resin obtained as a residue in the distillation of crude turpentine from the sap of the pine tree (gum rosin) or from an extract of the stumps and other parts of the tree (wood rosin).

Scarf Joint—See *Joint, Scarf.*

Self-Curing—See *Self-Vulcanizing.*

Self-Vulcanizing—Pertaining to an adhesive which undergoes vulcanization without the application of heat.

Separate Application Adhesive—See *Adhesive, Separate Application.*

Set—To convert an adhesive into a fixed or hardened state by chemical or physical action, such as condensation, polymerization, oxidation, vulcanization, gelation, hydration, or evaporation of volatile constituents. (See also *Cure* and *Dry.*)

Shelf Life—See *Storage Life.*

Shortness—A qualitative term that describes an adhesive that does not string, cotton, or otherwise form filaments or threads during application.

Single Spread—See *Spread.*

Size—See *Sizing.*

Sizing—The process of applying a material on a surface to fill pores and thus reduce the absorption of the subsequently applied adhesive or coating or to otherwise modify the surface properties of the substrate to improve the adhesion. Also, the material used for this purpose. The latter is sometimes called *Size.*

Slippage—The movement of adherends with respect to each other during the bonding process.

Solids Content—The percentage by weight of the nonvolatile matter in an adhesive.

Specific Adhesion—See *Adhesion.*

Spread—The quantity of adhesive per unit joint area applied to an adherend. It is preferably expressed in pounds of liquid or solid adhesive per thousand square feet of joint area: (1) *Single Spread* refers to application of adhesive to only one adherend of a joint; (2) *Double Spread* refers to application of adhesive to both adherends of a joint.

Storage Life—The period of time during which a packaged adhesive can be stored under specified temperature conditions and remain suitable for use. Sometimes called *Shelf Life.* (See also *Working Life.*)

Starved Joint—See *Joint, Starved.*

Strength, Dry—The strength of an adhesive joint determined immediately after drying under specified conditions or after a period of conditioning in the standard laboratory atmosphere. (See also *Strength, Wet.*)

Strength, Wet—The strength of an adhesive joint determined immediately after removal from a liquid in which it has been immersed under specified conditions of time, temperature, and pressure.

Stringiness—The property of an adhesive that results in the formation of filaments or threads when adhesive transfer surfaces are separated. (See *Webbing.*)

Substrate—A material upon the surface of which an adhesive-containing substance is spread for any purpose, such as bonding or coating. A broader term than adherend. (See also *Adherend.*)

Syneresis—The exudation of small amounts of liquid by gels on standing.

Tack—Stickiness of an adhesive.

Tack, Dry—The property of certain adhesives, particularly nonvulcanizing rubber adhesives, to adhere on contact to themselves at a stage in the evaporation of volatile constituents, even though they seem dry to the touch. Sometimes called *Aggressive Tack.*

Tack-Range—The period of time in which an adhesive will remain in the tacky-dry condition after application to an adherend, under specified conditions of temperature and humidity.

Tacky-Dry—Pertaining to the condition of an adhesive after the volatile constituents are either evaporated or absorbed sufficiently to leave it in a desired tacky state.

Thermoplastic—Capable of being repeatedly softened by heat and hardened by cooling.

Thermoset—A material which will undergo or has undergone a chemical reaction by the action of heat, catalysts, ultraviolet light, etc., leading to a relatively infusible state.

Thermosetting—having the property of undergoing a chemical reaction by the action of heat, catalysts, ultraviolet light, etc., leading to a relatively infusible state.

Thinner—A volatile liquid added to an adhesive to modify the consistency or other properties. (See also *Diluent* and *Extender.*)

Thixotrophy—A property of adhesive systems which thin upon isothermal agitation and thicken upon subsequent rest.

Time, Assembly—The time interval between the spreading of the adhesive on the adherend and the application of pressure or heat, or both, to the assembly.

Time, Curing—The period of time during which an assembly is subjected to heat or pressure, or both, to cure the adhesive. (See also *Time, Joint Conditioning* and *Time, Setting.*)

Time, Drying—The period of time during which an adhesive on an adherend or an assembly is allowed to dry with or without the application of heat or pressure, or both. (See also *Time, Curing; Time, Joint Conditioning;* and *Time, Setting.*)

Time, Joint Conditioning—The time interval between the removal of the joint from the conditions of heat or pressure, or both, used to accomplish bonding and the attainment of approximately maximum bond strength. Sometimes called *Joint Aging Time.*

Time, Setting—The period of time during which an assembly is subjected to heat or pressure, or both, to set the adhesive. (See also *Time, Curing; Time, Joint Conditioning;* and *Time, Drying.*)

Viscosity—The internal frictional resistance of an adhesive to flow when that resistance is directly proportional to the applied force.

Viscosity Coefficient—The shearing stress tangentially applied which will induce a velocity gradient. A material has a viscosity of one poise when a shearing stress of one dyne per square centimeter produces a velocity gradient of $\frac{\text{1 cm per sec}}{\text{sec}}$ (See *Viscosity.*)

Warm-Setting Adhesive—See *Adhesive, Warm-Setting.*

Warp—A significant variation from the original, true or plane surface.

Webbing—Filaments or threads that may form when adhesive transfer surfaces are separated. (See *Stringiness.*)

Wet Strength—See *Strength, Wet.*

Wood, Built-Up Laminated—An assembly made by joining layers of lumber with mechanical fastenings so that the grain of all laminations is essentially parallel.

Wood, Glued Laminated—An assembly made by bonding layers of veneer or lumber with an adhesive so that the grain of all laminations is essentially parallel.

Wood Failure—The rupturing of wood fibers in strength tests on bonded specimens, usually expressed as the percentage of the total area involved which shows such failure.

Wood Veneer—A thin sheet of wood, generally within the thickness range of .01 to .25 in., to be used in a laminate.

Working Life—The period of time during which an adhesive, after mixing with catalyst, solvent, or other compounding ingredients, remains suitable for use. (See also *Storage Life.*)

Yield Value—The stress (either normal or shear) at which a marked increase in deformation occurs without an increase in load.

bibliography

Abolins, V., and Eickert, J. "Adhesive Bonding and Solvent Cementing of Polyphenylene Oxide," *Adhesives Age,* Vol. 10 (July, 1967), pp. 22–6.

Adhesion Fundamentals and Practice. Report of an International Conference held at the University of Nottingham, England, September 20–22, 1966. New York: Gordon and Breach Science Publishers, 1969.

Adhesive Bonding Alcoa Aluminum. Pittsburgh: Aluminum Company of America, 1967.

"Adhesive Bonding Holds Down Costs," *Metal Progress,* Vol. 93 (May, 1968), pp. 111–12.

"Adhesive—Bonding Hood Assemblies," *American Machinist/Metalworking Manufacturing,* August, 1962.

"Adhesive Replaces Welding in Pressure-Tight Assembly," *Adhesives Age,* Vol. 11 (May, 1968), p. 24.

"Adhesive Simplifies Coating Application," *Adhesives Age,* Vol. 10 (July, 1967), p. 39.

"Adhesive Tester Permits Multiple Specimen Reports," *Adhesives Age,* Vol. 10 (March, 1967), p. 37.

"Adhesive Weights and Balance Wheels," *Adhesives Age,* Vol. 10 (January, 1967), p. 27.

"Air Conditioner Quality Improved through Use of Adhesives and Sealers," *Adhesives Age,* Vol. 8, No. 9 (September, 1965), pp. 38–9.

"Air-Powered Applicators Double Plant Production," *Adhesives Age,* Vol. 8, No. 4 (April, 1965), pp. 26–7.

Alexander, M., ed. "The Case for Metal to Metal Adhesive Bonding," *Production,* Vol. 57, No. 5 (May, 1966), pp. 112–23.

Alner, D. J. *Aspects of Adhesion.* vol.'s 1, 2 and 3. England: University of London Press, 1965, 1966 and 1967.

Alter, H. "Filler Particle Size and Mechanical Properties of Polymers," *Journal of Applied Polymer Science,* Vol. 9, No. 4 (April, 1965), pp. 1525–31.

Anderson, C. C. "Adhesives—Annual Review," *Industrial and Engineering Chemistry,* Vol. 59 (August, 1967), pp. 91–6.

Baldwin, F. P.; Fusco, J. V.; and Gastwirt, L. E. "Elastomeric Prepolymers for Adhesives and Sealants Provide Improved Strength and Versatility," *Adhesives Age,* Vol. 10, No. 2 (February, 1967), pp. 22–9.

Barbarisi, M. J. *Wetting of High and Low Energy Surfaces by Liquid Adhesives and Its Correlation to Bond Strength,* Bulletin No. AD-651 093. Dover, N.J.: Picatinny Arsenal, 1967.

Barenholtz, G. J.; Rutzler, J. E., Jr.; and Bobalik, E. G. "Correlation of Some Thermal Properties of Epoxy Adhesives and Shear Strength of Steel Bonds," *Industrial and Engineering Chemistry—Product Research and Development,* Vol. 2, No. 4 (December, 1963), pp. 270–2.

Barrett, R. R., and Lawrence, R. D. "Factors Affecting Phasing in Neoprene Solvent Adhesives," *Adhesives Age,* Vol. 9, No. 10 (October, 1966), pp. 22–6.

Bartlett, F. P. "Adhesives for Films and Foils," *Adhesives Age,* Vol. 9, No. 2 (February, 1966), pp. 34–5.

Beaver, W. "Save Space and Cut Costs with Bulk Adhesive Handling," *Adhesives Age,* Vol. 8, No. 4 (April, 1965), p. 38.

Bertozzi, E. R. "Chemistry and Technology of Elastomeric Polysulfide Polymers," *Rubber Chemistry and Technology,* Vol. 41 (February, 1968), pp. 114–60.

"Better Adhesion to Plastic Bottles," *Adhesives Age,* Vol. 9, No. 3 (March, 1966), pp. 32–3.

Bikerman, J. J. *The Science of Adhesive Joints.* New York: Academic Press, 1961.

______. "Theory of Adhesion for Joint Sealants," *National Research Council—Building Research Institute—New Joint Sealants*—Publication 1006 (1963), pp. 35–9.

______. "Testing for Wetting Properties of Plastics Surfaces," *Testing of Polymers,* Vol. 2 (1966), pp. 135–49.

______. "Causes of Poor Adhesion; Weak Boundry Layers," *Industrial and Engineering Chemistry,* Vol. 59 (September, 1967), pp. 40–4.

Billmeyer, F. W., Jr. *Textbook of Polymer Science.* New York: Interscience Publishing Company, 1962.

Black, J. M., and Blomquist, R. F. "Metal-Bonding Adhesives for High-Temperature Service," *Modern Plastics,* Vol. 33, No. 10 (June, 1956), pp. 225–8, 230, 235–6.

Blomquist, R. F. "Adhesives—Past, Present and Future," *ASTM—Special Technical Publication No. 360* (1964), pp. 179–212.

Bodnar, M. J., ed. *Applied Polymer Symposia No. 3: Structural Adhesive Bonding.* New York: Interscience Publishing Company, 1966.

"Bonding De-icer Boats to Helicopter Blades," *Adhesives Age,* Vol. 8, No. 7 (July, 1965), p. 29.

"Bonding of Wet Suits Reduces Repair Costs," *Adhesives Age,* Vol. 10, No. 2 (February, 1967), p. 38.

"Bonding Soundproofing Pads to Folding Steel-Door Panels," *Adhesives Age,* Vol. 8, No. 11 (November, 1965), p. 29.

"Bonding Vinyl Carpeting Cuts Warranty Costs for Builder of Small Boats," *Adhesives Age,* Vol. 8, No. 4 (April, 1965), p. 25.

Booth, C. C., and Duncan, T. F. "Elastomeric Adhesives: Characteristics and Applications," *Adhesives Age,* Vol. 4, No. 7 (July, 1961), pp. 34–8.

Brams, S. L. "Factors Involved in Manufacturing Rubber Adhesives," *Adhesives Age,* Vol. 10 (March, 1967), pp. 32–7.

Braude, F. *Adhesives.* New York: Chemical Publishing Company, Inc., 1943.

Brewis, D. M. "Surface Properties Associated with Adhesion," *SPE—2nd Annual Technical Conference,* Vol. 12, *Paper X-1* (March, 1966), p. 3.

Bugel, T. E.; Norwalk, R.; and Snedeker, R. H. "Phenoxy Resin—New Thermoplastic Adhesive," *ASTM—Special Technical Publication No. 360* (1964), pp. 87–95.

"Building Panels Strengthened with Wide-Range Adhesive," *Adhesives Age,* Vol. 8, No. 5 (May, 1965), pp. 26–7.

Burgman, H. A. "Selecting Structural Adhesive Materials," *Electro-Technology,* Vol. 75, No. 6 (June, 1965), pp. 69–75.

———, et al. "Aromatic Imide Polymers for High-Temperature Adhesives," *Journal of Applied Polymer Science,* Vol. 12 (April, 1968), pp. 805–829.

Cagle, C. V. *Adhesive Bonding: Techniques and Applications.* New York: McGraw-Hill Book Company, 1968.

Chessin, N., and Taylor, G. "Working Life of Room-Temperature-Curing Epoxy Adhesives," *Adhesives Age,* Vol. 10 (September, 1967), pp. 29–31.

Clarke, J. A., and Hawkins, J. M. "Characteristics of a Urethane Modified Structural Epoxy Adhesive," *Preprints of Division of Organic Coating and Plastics Chemistry,* Vol. 28, No. 1. American Chemical Society 155th Meeting (April, 1968), pp. 468–75.

Clarke, J. F., et al. "Review of Ceramics—Metal Joining," AFML-TR-65-143 (May, 1965).

Colovos, G. C.; McClellan, T. R.; and Rausch, K. W. "Polyurethane Sealants—Some Improved Systems," *Adhesives Age,* Vol. 9, No. 11 (November, 1966), pp. 23–7.

"Contact Adhesive Cuts Cabinet Door Costs," *Adhesives Age,* Vol. 10 (March, 1967), p. 31.

Conway, D. J. "Metal-Metal Adhesives for Structural Applications at Elevated Temperatures," *Adhesives Age,* Vol. 11 (May, 1968) p. 30.

Cooke, H. G.; McWhorter, W. F.; and Strohscher, A. R. "Chemical Resistance of Epoxy Resins," *Industrial and Engineering Chemistry,* Vol. 56, No. 5 (May, 1964), pp. 38–41.

Coppel, J. M., and Enthoven, P. H. "Epoxides in Constructional Industry," *Plastics,* Vol. 30, No. 328 (February, 1965), pp. 82–4.

Dahlquist, C. A. "Adhesion and Cleavage Stress at Interface between Solids and Pressure-Sensitive Adhesive," *ASTM—Special Technical Publication No. 360* (1964), pp. 46–58.

Damusis, A. *Sealants.* New York: Reinhold Publishing Company, 1967.

Dannenberg, H. "Determination of Stresses in Cured Epoxy Resins," *SPE—Journal,* Vol. 21 (July, 1965), pp. 669–95.

Davis, I. J. "Vinyl-to-Metal Laminating," *Adhesives Age,* Vol. 3, No. 3 (March, 1960), pp. 40–2.

Dawins, B.; Hodgson, S.; and Tasker, L. "Water Borne Adhesives Based on Polyvinyl Acetate," *Rubber and Plastics Age,* Vol. 47, No. 3 (March, 1966), pp. 289–291–2.

Debnar, D. "Adhesives Used in Packaging and Converting Industries," *Adhesives Age,* Vol. 9, No. 2 (February, 1966), pp. 30–1.

Debruyne, N. A., and Houwink, R. *Adhesion and Adhesives,* Chapter 5C. New York: Elsevier Publishing Company, 1951.

Delmonte, J. "Bonding of Systems," *SPE—25th Annual Technical Conference,* Vol. 13 (May 15–18, 1967), pp. 132–3.

———. "Bonding Thermoplastics with Improved Urethanes," *Adhesives Age,* Vol. 11 (April, 1968), pp. 27–8.

———. "Solventless Bonding of Polyvinyl Resins," *SPE—25th Annual Tech. Conference,* Vol. 13 (May 15–18, 1967), pp. 132–3.

———. *Technology of Adhesives.* New York: Reinhold Publishing Company, 1947.

DePalma, E. P. "Topic No. 25," (Paper for ASM Metals Handbook Committee on Adhesive Bonding of Steel, American Society of Metals, September 14, 1964).

"Do Hot Melts Need an Apology?" *The Thermogrip Applicator,* No. 8. USM Corporation (October, 1966), pp. 2–3.

Dombrow, B. A. *Polyurethanes.* 2nd ed. New York: Reinhold Publishing Company, 1965.

Dow Epoxy Resins. Dow Chemical Company Technical Bulletin No. 170-188.

Dow Liquid Epoxy Resins. Dow Chemical Company Technical Bulletin No. 170-141A.

Drisko, R. W.; Alumbaugh, R. L.; and Cobb, J. W. "Testing Adhesion of Epoxies that Cure Underwater," *Materials Protection,* Vol. 4, No. 9 (September, 1965), pp. 24–8.

DuPont Company Technical Newsletter No. 1. "Lamination with Surlyn A, Ionomer Resin," Bulletin No. A-50740, August, 1966.

Edelman, I. L.; Barabash, M. L.; and Glovatskaya, E. P. "Adhesion to Metals of Polymer Films with Fine Particle Fillers," *Plasticheskie Massy,* No. 1 (January, 1966), pp. 62–4.

Einhorn, S. C., and Turetsky, S. B. "Rheological Properties of SBR Polymers by Capillary Extrusion," *Journal of Applied Polymer Science,* Vol. 8, No. 3 (May, 1964), pp. 1257–73.

"Electronic Heat-Sealing Keeps Up to Date," *Modern Plastics,* Vol. 44 (July, 1967), pp. 86–9, 148.

Eley, D. D., ed. *Adhesion.* Oxford University Press, 1961.

Epstein, G. *Adhesive Bonding of Metals.* New York: Reinhold Publishing Company, 1954.

"Equivalence of Cyclohexanone and Tetrahydrofuran as Solvents for Polyvinyl Chloride," *Journal of Applied Polymer Science,* Vol. 9, No. 6 (June, 1965), pp. 2319–20.

Floyd, D. E. *Polyamide Resins.* 2nd ed. New York: Reinhold Publishing Company, 1966.

"Foam Adhesive Speeds Production," *Adhesives Age,* Vol. 10, No. 6 (June, 1967), p. 35.

Gardner, E. D., and Flournoy, R. W. "Adhesive Bonding of Aluminum," *Adhesives Age,* Vol. 4, No. 1 (January, 1961), pp. 20–5.

Gaylord, N. G., and Mark, H. *Linear and Stereoregular Addition Polymers.* New York: Interscience Publishing Company, 1959.

"Glued Walls for New Building," *Adhesives Age,* Vol. 8, No. 2 (February, 1965), pp. 36–7.

Goldman, E. J.; Rosenberg, R. R.; and Lee, W. E. "How to Use Fluorocarbon Plastics as Bonding Agents," *Adhesives Age,* Vol. 10 (October, 1966), pp. 22–6.

Gorton, B. S. "Interaction of Nylon Polymers with Epoxy Resins in Adhesive Blends," *Journal of Applied Polymer Science,* Vol. 8, No. 3 (May, 1964), pp. 1287–95.

______. "Adhesive Properties of Polyisophthalamides," *Journal of Applied Polymer Science,* Vol. 9, No. 11 (November, 1965), pp. 3753–8.

Gray, V. R. "Surface Aspects of Wetting and Adhesion," *Chemistry and Industry,* No. 23 (June, 1965), pp. 969–78.

Guttman, W. H. *Concise Guide to Structural Adhesives.* New York: Reinhold Publishing Company, 1961.

Guziak, L. F., and Maclay, W. N. "Redispersible Latex Polymer," *Journal of Applied Polymer Science,* Vol. 7, No. 6 (November, 1963), pp. 2249–58.

Hansen, R. H., and Schonhorn, H. "A New Technique for Preparing Low Surface Energy Polymers for Adhesive Bonding," *Journal of Polymer Science*—Polymer Letters Part B, Vol. 4, No. 3 (March, 1966), pp. 203–9.

Hardman, K. "Packaging and Utilization of Two-Part Epoxies," *Adhesives Age,* Vol. 10 (September, 1967), pp. 26–8.

Head, J. K. "Doubling Adhesives," *Adhesives Age,* Vol. 8, No. 3 (March, 1965), pp. 32–4.

Henderson, A. W. "Introduction to Adhesives," *Sheet Metal Industries,* Vol. 42, No. 461 (September, 1965), pp. 653–61.

Hess, E. F. "Adhesives for Fabricating High Strength Structures," *Metal Progress,* Vol. 83 (June, 1963), pp. 97–100, 120–8.

"High-Strength Metal Adhesives," *Adhesives Age,* Vol. 10 (January, 1967), p. 10.

Hiler, M. J. "Adhesive Failures; Reasons and Preventions," *Adhesives Age,* Vol. 10 (January, 1967), pp. 28–31.

Hill, J. R. "Process Development of Polybenzimidazole Adhesives," *Adhesives Age,* Vol. 9, No. 8 (August, 1966), pp. 32–6.

Hinterwaldner, R. "Verbinden von Bauteilen mit Epoxydharzklebern," *Adheasion,* No. 1 (January, 1966), pp. 5–11.

Hock, C. W. "Pressure-Sensitive Adhesives—What Makes Them Stick," *Adhesives Age,* Vol. 7, No. 3 (March, 1964), pp. 21–5.

"Hot Melts; Types of Adhesives Used in Electrostatic Flocking," *Modern Textiles Magazine,* Vol. 49 (June, 1968), pp. 31–2.

"Hot-Spray Adhesives Speed Production of Sandwich Panels," *Adhesives Age,* Vol. 10, No. 5 (May, 1967), pp. 30–1.

Houwink, R., and Solomon, G., eds. *Adhesion and Adhesives,* Vol. 2, 2nd Edition. New York: Elsevier Publishing Company, 1967.

"How Adhesives Are Used at a Plywood Facility," *Adhesives Age,* Vol. 10 (June, 1967), pp. 36–7.

"How Infrared Ovens Speed Curing of Adhesives on Carpet Backs," *Adhesives Age,* Vol. 10 (July, 1967), pp. 38–9.

"How Oil Company Reduces Maintenance Costs with Resin Adhesive," *Adhesives Age,* Vol. 10, No. 4 (April, 1967), pp. 28–9.

"How to Prevent Solvent Hazards in Adhesive Plants," *Adhesives Age,* Vol. 10 (July, 1967), pp. 34–5.

Howe, H. E., and Radke, S. F. "How to Join Lead with Adhesives," *Adhesives Age,* Vol. 11 (September, 1968), pp. 22–31.

Huff, L. A. "Acrylic Aqueous and Solvent Base Caulks and Sealants," *Adhesives Age,* Vol. 10 (May, 1967), pp. 27–9.

Huntsberger, J. R. "Mechanisms of Adhesion," *Journal of Paint Technology,* Vol. 39, No. 507 (April, 1967), pp. 199–211.

Irving, R. R. "Adhesives Bonding; Its Day is Coming," *Iron Age,* Vol. 201 (April 11, 1968), pp. 83–90.

Jones, D. L., and DePalma, E. P. "A Tailor-Made High Product Metal to Metal Automotive Adhesive," *SAE Technical Paper,* March, 1960.

Kalb W. J. "Ford Sticks to Glued-on Trimmings," *Iron Age,* Vol. 201 (March 21, 1968), p. 21.

Kaliske, G. "Herstullung von Metallklebuerbindungen mit Hilfe von Teer-Epoxidharzen," *Plaste Kautschuk,* Vol. 12, No. 8 (August, 1965), pp. 453–55.

Kanamaru, K. "Rheological Treatment of Rate of Setting of Adhesive Joints," *Kolloid-Zeit,* Vol. 192, No.'s 1–2 (October, 1963), pp. 51–66.

Kantner, G. C. "Adhesives for Textiles; Discussing Preparation and Testing of Laminates and Flocked Fabrics," *Adhesives Age,* Vol. 10 (June, 1967), pp. 22–9.

Katz, I. *Adhesive Materials: Their Properties and Usage.* Long Beach, Calif.: Foster Publishing Company, 1964.

———. "Guides to Military and Federal Specifications for Rubber-Bonding Adhesives," *Adhesives Age,* Vol. 8, No. 2 (February, 1965), pp. 24–30.

Kaufman, J. G. "Fracture Toughness of 7075-T6 and T651 Sheet, Plate, and Multilayered Adhesive—Bonded Panels," *Journal of Basic Engineering* (September, 1967), pp. 503–7.

Kausen, R. C. *High and Low-Temperature Adhesives.* Material Symposium, Bulletin No. CA62:2884a. Los Angeles, California: National Society of Aerospace Material Process Engineers, 1964.

———. "Adhesives for High and Low Temperatures—I," *Materials in Design Engineering,* Vol. 60 (August, 1964), pp. 94–8.

Klinetob, A. B. "Proper Care of Spreader Reduces Gluing Problems," *Adhesives Age,* Vol. 9, No. 1 (January, 1966), p. 25.

Knight, R. *Adhesives for Wood.* New York: Chemical Company, Inc., 1952.

Kreibich, R. E., and Freeman, H. G. "Testing Adhesives for Creep," *Adhesives Age,* Vol. 8, No. 8 (August, 1965), pp. 29–34.

Lambert, J. M. "How a Glue Gun System Solves Adhesive and Sealant Problems," *Adhesives Age,* Vol. 10 (May, 1967) pp. 24–6.

"Latex Adhesive Effectively Bonds Insulation to Metal Panels," *Adhesives Age,* Vol. 11 (March, 1968), p. 31.

Lauren, S. "Adhesives for Wallboard," *Adhesives Age,* Vol. 8, No. 5 (May, 1965), pp. 30–1.

Lawrence, J. R. *Polymer Resins.* New York: Reinhold Publishing Company, 1960.

Lee, H., and Neville, K. *Handbook of Epoxy Resins.* 2nd ed. New York: McGraw-Hill Book Co., 1967.

Lewis, A. F. "Postformable Adhesively Bonded Metal Laminates," *Adhesives Age,* Vol. 10 (May, 1967), pp. 20–3.

Leonard, F. "Synthesis and Degradation of Poly-Cyanoacrylates," *Journal of Applied Polymer Science,* Vol. 10, No. 2 (February, 1966), pp. 259–72.

Livtak, S. "Polybenzimidazole Adhesives for Bonding Stainless Steel, Beryllium and Titanium Alloys. Part I: Adhesive Performance of PBI," *Adhesives Age,* Vol. 11 (January, 1968), pp. 17–24.

———. "Polybenzimidazole Adhesives for Bonding Stainless Steel, Beryllium and Titanium Alloys. Part II: Adhesive Sandwich Study," *Adhesives Age,* Vol. 11 (February, 1968), pp. 24–28.

Loan, L. D. "Natural Latex Adhesives," *Adhesives Age,* Vol. 8, No. 3 (March, 1965), p. 36.

Mackay, H. A. "Evaluation of Polybenzimidazole Glass-Fabric Laminates," *Modern Plastics,* Vol. 43 (January–April, 1966), pp. 149–247.

Martin, R. A. "Neoprene Phenolic Adhesives," *Adhesives Age,* Vol. 9, No. 5 (May, 1966), pp. 28–30.

———, and Traver, C. R. "Resins in Hot-Melt Coatings," *Tappi,* Vol. 49, No. 10 (October, 1966), pp. 51A–3A.

Matthew, C. "Break-Through in Hot Melt Adhesives," *Production*, Vol. 40, No. 11 (November, 1964), pp. 51–2.

McCarter, L. H. "Patents and Adhesives; How Patents Affect the Adhesives Field," *Adhesives Age*, Vol. 10 (May, 1967), pp. 32–4.

McGuire, E. P. *American Adhesives Index*. New York: Padric Publishing Company, 1962.

______. *Packaging and Paper Converting Adhesives*. New York: Palmerton Publishing Company, 1963.

______. *Adhesive Raw Materials Handbook*. Bernardville, N.J.: Padric Publishing Company, 1967.

Military Standardization Handbook; Adhesives, MIL-HDBK-691A. Washington: Department of Defense, 1965.

Miller, M. L. *Structure of Polymers*. New York: Reinhold Publishing Corporation, 1966.

Minnerly, H. E. "Processibility of Nitrile Polymers," *Rubber World*, Vol. 152, No. 1 (April, 1965), pp. 76–9.

Mittrop, F. "Das Verbinden von Kupfer und Messing mit Kunstharzklebstoffen, *Z. Metallk.*, Vol. 56, No. 9 (September, 1965), pp. 622–5.

"Multiwall with Hot-Melt Seal," *Modern Packaging*, Vol. 40, No. 3 (November, 1966), pp. 116–17.

Murray, E. J. "Adhesives and Adhesion," *Adhesives Age*, Vol. 8, No. 4 (April, 1965), pp. 23–4.

______. "Raw Materials and Adhesives for Paper Bonding," *Adhesives Age*, Vol. 10 (September, 1967), pp. 22–5.

"Nameplate Manufacturer Improves Performance with a Film Adhesive," *Adhesives Age*, Vol. 10 (June, 1967), pp. 34–5.

Neuss, H., ed. "Testing of Adhesives," *Tappi Monograph* No. 26. New York: Technical Association Pulp and Paper Industry, 1963.

"New Adhesive System Saves Time, Money," *Chemistry and Engineering News*, Vol. 46, No. 18 (April 22, 1968), pp. 40–1.

"Nondestructive Testing of Adhesive Bonds in Aircraft," *Adhesives Age*, Vol. 10 (August, 1967), pp. 35–7.

"Packaging Adhesive Solves Cellophane Window Problems," *Adhesives Age*, Vol. 10 (June, 1967), p. 29.

Palin, G. R. *Plastics for Engineers*. Oxford, England: Pergamon Press, Inc., 1967.

Parker, R. S. R., and Taylor, P. *Adhesion and Adhesives*. New York: Pergamon Press, 1966.

Pascuzzi, B., and Hill, J. R. "Structural Adhesives for Cryogenic Applications," *Adhesives Age*, Vol. 8, No. 3 (March, 1965), pp. 19–26.

Patrick, R. L., ed. *Treatise on Adhesives and Adhesions*, Vol. 1. New York: Marcel Dekker, 1967.

Pearson, J. R. A. *Mechanical Principles of Polymer Melt Processing*. Oxford, England: Pergamon Press, Inc., 1966.

Perry, H. A. *Adhesive Bonding of Reinforced Plastics*. New York: McGraw-Hill Book Company, Inc. 1959.

Perry, T. D. *Modern Wood Adhesives*. New York: Pitman Publishing Corporation, 1944.

Peterson, C. S. "Sealant Jungle," *Adhesives Age*, Vol. 10 (June, 1967), pp. 30–3.

Picotte, G. L. "Toughness of Structural Adhesives," *Adhesives Age*, Vol. 8, No. 2 (February, 1965), pp. 21–3.

Polniaszek, M. D. "How Silicone Primer Promotes Adhesion of Cable Clamps Sealant," *Adhesives Age*, Vol. 9, No. 11 (November, 1966), pp. 34–5.

Preiss, D. M. "Bonding Polyurethane to Metal," *Adhesives Age*, Vol. 9, No. 4 (April, 1966), pp. 22–4.

"Preparation, Circulation and Storage of Corrugating Adhesives, *Tappi*, 1965.

Proskauer, E. S., et al. *Perspectives in Polymer Science*. Interscience Publishing Company, 1966.

Radtki, S. F. "Adhesive Bonding Pb and Zn Opens New Uses; Cuts Costs," *Metals Engineering*, Vol. 67 (January, 1968), pp. 62–3.

Rayner, C. A. A. *Adhesion and Adhesives*. vol's 1 and 2. Amsterdam: Elsevier Publishing Company, edited by Houwink, R., and Salomon, G., 1965.

Rechter, H. L., and Haroda, Y. "High Temperature Strain Gage Adhesives," *Adhesives Age*, Vol. 8, No. 8 (March, 1965), pp. 27–31.

Reegen, S. L. "Peel Strength of Polyurethanes from Oxypropylene Polyols—I: Polyol Structure and Fillers," *Journal of Applied Polymer Science*, Vol. 9, No. 1 (January, 1965), pp. 279–94.

______. "Peel Strength of Polyurethanes from Oxypropylene Polyols—II: Effect of Additives, Isocyanate Structure, and Other Factors," *Journal of Applied Polymer Science*, Vol. 10, No. 9 (September, 1966), pp. 1247–59.

Reinhart, F. W., and Callomon, I. G. "Survey of Adhesion and Adhesives," *WADC Technical Report No. 58-450*. Ohio: Wright Air Development Center, 1959.

"Releasing Paper Provides Solution to Adhesive-Back Match Books," *Adhesives Age*, Vol. 10 (March, 1967), p. 40.

Richards, T. M., and Weltzin, R. D. "Adhesive Strengthens Riveted Assembly," *Adhesives Age*, Vol. 8, No. 12 (December, 1965), p. 35.

Rieke, J. K. "New Adherent Thermoplastic Polyethylenic Material—High Density Polyethylene Acrylic Acid Graft Copolymers," *SPE—22nd Annual Technical Conference*, Vol. 12 (March, 1966), p. 6.

Riley, M. W. "Adhesive Bonding—State of Art," *Materials in Design Engineering*, Vol. 54, No. 7 (December, 1961), pp. 89–93.

Robinson, H. A. "Hot Melts," *Assembly Engineering*, Vol. 8, No. 2 (February, 1965), pp. 43–7.

Roseland, L. M. "Structural Adhesives in Space Applications," *Applied Polymer Symposia No. 3* (1966), pp. 361–7.

———. "Developing Structural Adhesives Used for Space Applications," *Adhesives Age*, Vol. 9, No. 4 (April, 1966), pp. 32–5.

Sacharow, S. "Adhesive Needs of Flexible Packager," *Adhesives Age*, Vol. 8, No. 10 (October, 1965), pp. 26–8.

———. "Future of Adhesive Ink Formulations," *Adhesives Age*, Vol. 10 (July, 1967), p. 10.

———. "Food Packaging Adhesives of the Future," *Adhesives Age*, Vol. 10 (September, 1967), p. 8.

———. "Advances in Adhesive Drying," *Adhesives Age*, Vol. 11 (May, 1968), p. 8.

———. "ABC's of Foil Lamination," *Adhesives Age*, Vol. 11 (August–September, 1968), p. 8.

———. "Polyethylene Powder: a Unique Adhesive," *Adhesives Age*, Vol. 11 (October, 1968), p. 8.

Sandler, H. R., and Berg, F. R. "Polyurethanes as Cryogenic Adhesives," *Journal of Applied Polymer Science*, Vol. 9, No. 12 (December, 1965), pp. 3909–16.

Sawyer, J. W., and Stuart, R. E. "Ionomer, Laminates—New Plastic/Metal Composites," *Modern Plastics*, Vol. 44, No. 10 (June, 1967), pp. 125–8.

Scales, M. "Epoxy Based Structural Adhesives," *Adhesives Age*, Vol. 7, No. 11 (Nov., 1964), pp. 22–4.

Schmidt, D. L., and Jones, W. C. "Carbon Fiber Plastic Composites," *SPE—Journal*, Vol. 20, No. 2 (February, 1964), pp. 162–69.

Schmutzler, G., "Liquid Thioplastics and Their Application Possibilities," *Technik*, Vol. 19, No. 8 (1964), pp. 530–32.

Schneider, H. "Evaluation of Strippers for Epoxy Adhesives," *Plastics Design and Processing*, Vol. 5, No. 7 (July, 1965), pp. 18–21.

Schonhorn, H., and Sharpe, L. H. "What Makes Adhesive Stick," *Iron Age*, Vol. 194, No. 12 (September 17, 1964), pp. 84–6.

———, and ———. "Surface Adhesive Joints," *Journal of Polymer Science—General Papers Part A*, Vol. 3, No. 9 (September, 1965), pp. 3087–97.

"Scotch-Weld Brand Structural Adhesives," *3M Company*, St. Paul, Minnesota, 1967.

Selbo, M. L. "Adhesives, Their Use and Performance in Structural Lumber Products," *U. S. Forest Products Laboratory*—Report No. 2199 (September, 1960).

"Seminars on Adhesives for Industry," Madison, Wisconsin, *Adhesives Age*, Vol. 10 (April, 1967), pp. 37–8.

Sharpe, Louis H. "Assembling with Adhesives," *Machine Design*, Aug. 18, 1966.

Sharpe, L. H.; Schonhorn, H.; and Lynch, C. J. "Adhesives," *International Science and Technology*, No. 28 (April, 1964), pp. 26–37.

Skeist, I., ed. *Handbook of Adhesives*. New York: Reinhold Publishing Company, 1962.

———. *Encyclopedia of Polymer Science and Technology*. Vol. 1. New York: John Wiley & Sons, Inc., 1964.

Smarook, W. H., and Bonotto, S. "Adhesion of Carboxyl—Containing Polyolefins," *Journal of Polymer Engineering and Journal of Polymer Science*, Vol. 8 (January, 1968), pp. 41–49.

———, and ———. "Adhesion of Carboxyl Containing Polyolefins," *SPE—25th Annual Technical Conference*, Vol. 13 (May 15–18, 1967), pp. 119–31.

Smith, D. R. "How to Prepare Surface of Metals for Adhesive Bonding," *Adhesives Age*, Vol. 10, No. 3 (March, 1967), pp. 25–31.

Smith, P. S. *Synthetic Adhesives*. New York: Chemical Publishing Company, 1943.

Smith, W. M. *Vinyl Resins*. New York: Reinhold Publishing Company, 1958.

"Specially Equipped Spreaders Save Time and Labor in Timber Lamination," *Adhesives Age*, Vol. 9, No. 3 (March, 1966), p. 27.

"Sprayed Elastomer Costs," *Adhesives Age*, Vol. 8, No. 2 (February, 1965), p. 35.

Steel Structures Painting Manual. Pittsburgh: Steel Structures Painting Council, 1955.

Sterman, S., and Toogood, J. B. "How to Promote Adhesion with Silicones and Silanes," *Adhesives Age*, Vol. 8, No. 7 (July, 1965), pp. 34–9.

Strauss, E. L. "Strength of Polybenzimidazole and Phenolic Laminate-to-Metal Joints," *Polymer Engineering and Science*, Vol. 6, No. 1 (January, 1966), pp. 24–9.

Structural Adhesives: The Theory and Practice of Gluing with Synthetic Resins. London: Lange, Maxwell and Springer, 1952.

"Structural Sandwich Constructions; Wood; Adhesives," *Book of ASTM Standards,* Part 16. Philadelphia: American Society for Testing and Materials, 1968.

"Superstrength Adhesives Reduce Number of Holes in Car Bodies," *Product Engineering,* Vol. 39 (April 22, 1968), p. 50.

Swanson, F. "How to Develop and Use an Adhesive Evaluation System," *Adhesives Age,* Vol. 10 (April, 1967), pp. 30–4.

"Synthetic and Protein Adhesives for Paper Coating," *Tappi,* 1961.

Tanner, W. C., and Wegman, R. F. "Effects upon Strength of Epoxy Adhesives When Stressed to Failure in Milliseconds," *Adhesives Age,* Vol. 8, No. 9 (September, 1965), pp. 32–7.

Thelen, E. "Adherend Surface Preparation," *Journal of Applied Polymer Science,* Vol. 6, No. 20 (March–April, 1962), pp. 150–4.

Tobolsky, A. V., and Rembaum, R. "New Class of Block Polymers," *Journal of Applied Science,* Vol. 8, No. 1 (January, 1964), pp. 307–24.

Traxler, R. N. *Asphalt.* New York: Reinhold Publishing Corporation, 1961.

Truax, D. E. "Hot Melt Adhesives and Their Uses," *Tappi,* Vol. 49 (May, 1966), pp. 100A–2A.

Twiss, S. B. "Structural Adhesive Bonding—1, 2," *Adhesives Age,* Vol. 8, No. 1 (January, 1965), pp. 30–4.

______. "Preview of Mechanisms and Improvements in Adhesives of Tomorrow," *Adhesives Age,* Vol. 9, No. 1 (January, 1966), pp. 16–21.

Uffner, M. W., and Friefeld, M. "Vinylpyrrolidone Copolymer Adhesives," *Adhesives Age,* Vol. 8, No. 1 (January, 1965), pp. 18–22.

U. S. Patent No. 3 372 144 (March 5, 1968).

U. S. Patent No. 2 829 984 (April, 1968).

"Using Adhesive for Press Fitted Parts," *Adhesives Age,* Vol. 10 (July, 1967), p. 26.

Vasenin, R. M. "Adhesion of High Polymers," *Adhesives Age,* Vol. 8, No. 5 (May, 1965), pp. 18–21.

Vertnik, L. R. "Bonding Wood with Polyamide Adhesives," *Adhesives Age,* Vol. 10, No. 8 (July, 1967), pp. 27–9.

Voyutskii, S. S. *Autohesion and Adhesion of High Polymers.* New York: Interscience Publishing Company, 1963.

______, et al. "How Temperature Dependence and Activation Energy Affects Adhesion of Polymers to Metals," *Adhesives Age,* Vol. 8, No. 11 (November, 1965), pp. 24–8.

Wasserman, A., ed. *Size and Shape of Contractile Polymers: Conversion of Chemical and Mechanical Energy.* Oxford, England: Pergamon Press, Inc., 1960.

Weggemans, D. M. "Contactlijmen," *Plastica,* Vol. 19, No. 3 (March, 1966), pp. 102–5.

Wegman, R. F., and Dickinson, R. H. *An Evaluation of Several Structural Adhesives for Use at Elevated Temperatures,* Bulletin No. AD-468 336. Dover, N.J.: Picatinny Arsenal, 1965.

Weiss, P., ed. *Adhesion and Cohesion—Proceedings of the Symposium on Adhesion and Cohesion.* General Motors Research Laboratories, Warren, Michigan, 1961. Amsterdam: Elsevier Publishing Company, 1962.

______. "Polymer Concept in Adhesives," *Journal of Polymer Science—Polymer Symposia Part C,* No. 12 (1966), pp. 169–83.

Weitz, P. "Polyvinyl Methyl Ether—A Novel Route to Improved Adhesion." *Paper presented at Adhesive and Sealant Council Spring Seminar,* 1966.

Weymann, H. P. "Naval Stores in Rubber-Based Adhesives," *Adhesives Age,* Vol. 8, No. 7 (July, 1965), pp. 25–8.

______. "Developing Adhesive Resins for Neoprene Cements," *Adhesives Age,* Vol. 9, No. 9 (September, 1966), pp. 27–31.

Wetzel, F. H., and Alexander, B. B. "Modified Rosin Tackifiers in Adhesives Based on Synthetic Elastomers," *Adhesives Age,* Vol. 7, No. 1 (January, 1964), pp. 28–35.

Willard, R. S. *Adhesive Patents, 1955–1963.* New York: Padric Publishing Company, 1964.

Williamson, D. V. S. "Hot Melt Adhesives in Europe—Trends and Developments," *Adhesives Age,* Vol. 8, No. 8 (August, 1965), pp. 24–7.

Yeager, R. E. "Development of Adhesive Systems for Extreme Environments," *Applied Polymer Symposia,* No. 3 (1966), pp. 369–95.

Young, A. W. "Adhesives Rise to New Plateau with Better Materials, Tests," *Iron Age,* Vol. 196, No. 14 (September 30, 1965), pp. 102–5.

Yoshino, S. Y.; Nadler, M. A.; and Richter, D. H. "Fabrication Methods for PBI Adhesive Beryllium Sandwich Structures," *Adhesives Age,* Vol. 10 (August, 1967), pp. 26–34.

Zisman, W. A. "Influence of Constitution on Adhesion," *Industrial and Engineering Chemistry,* Vol. 55, No. 10 (October, 1963), pp. 18–38.

index

A

Abrasive blast cleaning, 92
Abrasive Cleaning, 90–95
  blast cleaning, 92
  centrifugal wheel method, 92
  dry blasting, 94
  sandblasting, 92–93
  wet blasting, 93–94
Abrasive process, 91–92
Abrasives for blast cleaning, 94–95
Acrylate diesters, 41
Acrylics, 24, 27
Adhesive application methods
  air spray, 64
  airless spray, 64
  brush, 63–64
  extrusion, 63
  flow, 63
  hand (manual), 75–76
    equipment cost, 75–76
  roller, 63
  trowel, 63–64
Adhesive bonding
  advantages, 46
  disadvantages, 46–47
Adhesive development history (*table*), 2
Adhesive primers, 125
Adhesive selection requirements, 59–60
Adhesive types
  chemically reactive types, 1–19
    catalytic plural components-chemical cure, 1–13
    catalytic plural components-moisture cure, 14
    heat activated (one-part system), 14–19
  delayed-tack adhesives, 35–36, 61
  evaporation or diffusion types, 20–27, 60–61
    solvent-based systems, 20–25
    water-based systems, 25–27
  film adhesives, 36–40, 61
    epoxies, 36–37
    nitriles, 37
    olefin copolymers, 37–39
    phenolics, 39
    polyamides, 39
    polyolefins, 39–40
    polyvinyl butyral, 40
    polyvinyl chloride (PVC), 40
  hot melt types (cooling), 27–35, 61
    asphalt, 34–35
    ceramics, 35
    olefin copolymers, 30–31
    phenoxies, 31
    polyamides, 31–32
    polyesters, 32–33
    polyethylene and polypropylene, 33
    polyvinyl acetate (PVA), 33
    polyvinyl butyral, 33–34
    polyvinyl ethers, 34
  pressure sensitive types, 40–41, 61
Adhesive "wettability," 90
Adhesives
  application conditions, 62
  post application cleanup, 62
Advantages of adhesive bonding, 46
Air entrapment, 49
Air spray equipment availability, 76–77
Air spray, 68, 115–117
Aircraft industry, bonding, 157–159
Airless spray, 68, 115–117
Airless spray equipment availability, 76–77
Aluminum anodizing, 97

American Society for Testing and Materials (ASTM) specifications, 127
Anaerobic adhesives, 41
Anchor pattern, 95
Application methods
  machine, 108–115
    air blade coater, 114
    bench roll coater, 108–109
    dip roll coater, 113–114
    engraved roll coater, 114
    floor mounted roll coater, 108–109
    flow or curtain coaters, 114–115
    "kiss" type roll coater, 110–111
    pressure or squeeze roll coaters, 111–112
    reverse roll coater, 112–113
    two-roll reverse roll coater, 112–113
  manual, 105–108
    brush, 105–106
    flow brush, 106
    glue guns, 107
    hand dipping, 107
    oil can, 106
    rollers, 107–108
    silk screen process, 106–107
    spatula, 106
    squeeze bottle, 106
Applicators
  hand, 85–86
  hot melt adhesives, 84
  specialty glue, 85
Asphalt, 34–35
Atmosphere control, 125–126
ASTM (*see* American Society for Testing and Materials)
ASTM D471, 141
ASTM D543, 141
ASTM D894-49, 131
ASTM D896-66, 141
ASTM D897, 138
ASTM D903, 138
ASTM D905-49, 132
ASTM D906, 138
ASTM D906-64, 133
ASTM D950-54, 138
ASTM D1002, 138
ASTM D1002-64, 133
ASTM D1062-51, 136
ASTM D1151-61, 138
ASTM D1344-57, 130
ASTM D1781-62, 135
ASTM D1876-61T, 134
ASTM D2094-62T, 130
ASTM D2095-62T, 129
ASTM D2182, 140
ASTM D2182-63T, 132, 133
ASTM D2293-64T, 134
ASTM D2294-64T, 134
ASTM D2339-65T, 134
Automotive industry, bonding, 143–147 155–157

B

Benzimidazoles, 15
Blast cleaning abrasives, 94–95
Blast cleaning grades, 93
Blasting
  dry, 94
    centrifugal wheel method, 92
    wet, 93–94
Blasting equipment, 95
"Blushing," 62
Boat building industry, bonding, 147–152
Bond integrity, 45
Bond inspection and testing, 126–141
Bond strength, 60–61
Bonding case histories
  aircraft industry, 157–159
  automotive industry, 143–147, 155–157
  boat building industry, 147–152
  food packaging industry, 152–155

C

Cartridges, disposable, 75
Case histories (*see* Bonding case histories)
Case sealing with spray equipment, 121
"CASING" concept, 41
Catalytic plural components—chemical cure, 1–13
  epoxy, 3
  phenolics, 4, 8
  polyesters, 8
  polysulfides, 9
  polyurethanes, 10
  resorcinol-formaldehyde, 8
  silicones, 12
Catalytic plural components—moisture cure, 14
  silicones, 14
  urethanes, 14
Centrifugal wheel blasting, 92
Ceramics, 35
Cermets, 35
Chemical cleaning, 96–101
  alkaline cleaning, 96
  conversion coating, 97
  degreasing, 96
  etching, 96
  surface preparation for acetals, 100–101
  surface preparation for aluminum, 96–97
  surface preparation for copper alloys, 98
  surface preparation for elastomers, 98–99
  surface preparation for fluorocarbons, 99–100
  surface preparation for metals, 89–90

Chemical cleaning (*Continued*)
surface preparation for mild steel, 98
surface preparation for nonmetallic materials, 90
surface preparation for polyethylene, 101
surface preparation for stainless steel, 97–98
Chemical compatibility, 45
Chemical reactive types, 60
Chemically reactive types, 1–19
catalytic plural components—chemical cure, 1–13
epoxy, 3
phenolics, 4
polyesters, 8
polysufides, 9
polyurethanes, 10
resorcinol-formaldehyde, 8
silicones, 12
catalytic plural components—moisture cure, 14
silicones, 14
urethanes, 14
heat activated (one-part system), 14–19
benzimidazoles, 15–16
epoxies, 16–17
nylon, 17–18
phenolics, 18
polyvinyl acetals, 18–19
rubber compounds, 19
urethanes, 19
Cleaning
abrasive, 90–96
blast cleaning, 92
centrifugal wheel method, 92
dry blasting, 94
sandblasting, 92–93
wet blasting, 93–94
alkaline, 96
chemical, 96–101
alkaline cleaning, 96
conversion coating, 97
degreasing, 96
etching, 96
surface preparation for acetals, 100–101
surface preparation for aluminum, 96–97
surface preparation for copper alloys, 98
surface preparation for elastomers, 98–99
surface preparation for fluorocarbons, 99–100
surface preparation for metals, 89–90
surface preparation for mild steel, 98
surface preparation for nonmetallic materials, 90
surface preparation for polyethylene, 101
surface preparation for stainless steel, 97–98
flame, 102
miscellaneous methods, 101–103
Cleaning acrylic plastics, solvents for, 103
Cleavage test, 136-137
Coaters
blade, 83
curtain, 83
roller, 83
(*See also* Roll coaters)
Coefficient of thermal expansion, 90
Cold-setting adhesives, 119
Containers
bulk considerations, 67
in-plant container handling, 67
size, 67
special, 67
transfer of, 67
Contaminants, 89–90
Contamination of adhesives, 123
Control, quality (*see* Quality control)
Conversion coating, 97
Copolymers, olefin, 30–31, 37
Corona discharge, 41
Corona discharge treatment, 39
Corrosion tests, environmental, 141
Cost
hot melt equipment, 122
manual equipment, 75–76
material, 67
packaging, 154
Cost comparison
conventional sealant (*table*), 151
polysulfide sealant (*table*), 151
Cross-lap specimen test, 130–131
Curing, 14, 120, 126
evaporation type, 60–61, 62
heat, 120
heat activated adhesives, 61–62
hot melt, 62
pressure, 120–121
room temperature, 61
structural adhesive, 61
Curing and pretreatment techniques, 41
Curtain coater, 74–75
Cyanoacrylate, 14

D

Degreasing, 96
Delayed-tack adhesives, 35–36, 61
Destructive tests, 128–141
cleavage, 136–137
environmental corrosion, 141
environmental temperature and humidity, 138–140

Destructive tests (*Continued*)
  impact, 138
  peel, 134–136
  sandwich structures, 141
  shear and lap shear, 132–134
  tensile, 129–131
DICY (*see* dicyandiamide)
Dicyandiamide (DICY), 16, 36
Dicyclohexyl phthalate, 36
Diphenyl phthalate, 36
Diphenylmethane diisocyanate (MDI), 23
Disadvantages of adhesive bonding, 46–47
Dispensers, measured repeatable shot, 78
Dry blasting, 94
Dust control, 94

E

Environmental corrosion tests, 41
Environmental temperature and humidity tests, 138–140
Epoxies, 3, 16–17, 36–37
Epoxy-alloy structural adhesives (*table*), 8
Epoxy adhesive, strength values (*table*), 1
Epoxy-glass fabric, 48
Equipment
  availability of low to medium viscosity multiple component, 81–83
  blasting, 95
  cost considerations, 66–68
  extrusion, 70–74
  high viscosity multiple component reactive resins, 78–81
  hot melt, 122
  low to medium viscosity multiple component reactive resins, 81–83
  spray application, 115–118
  ventilating, 68
Equipment application considerations (*table*), 66
Equipment availability, 76–86
Equipment compatibility with adhesive, 65
Equipment freeze-up, 65
Equipment life expectency, 65
Equipment selection guidelines, 64–65
E/AA (*see* Ethylene/acrylic acid)
E/EA (*see* Ethylene ethyl acrylate)
E/VA (*see* Ethylene vinyl acetate)
Etching, 96
Ethylene/acrylic acid (E/AA), 37–38
Ethylene-based adhesives, peel values (*table*), 30
Ethylene ethyl acrylate (E/EA), 30–31
Ethylene vinyl acetate (E/VA), 30
Evaporation or diffusion types, 20–27, 60–61
  solvent-based systems, 20–25
    acrylics, 24–25
    natural rubber, 20
    phenolics, 22
    reclaimed rubber, 20–21
    synthetic rubbers, 21–22
    urethanes, 22–23
    vinyl resins, 23–24
  water-based systems, 25–27
    acrylics, 26–27
    natural rubber, 26
    reclaimed rubber, 26
    synthetic rubbers, 26
    vinyl resins, 26
External mix nozzle, 118
Extrusion equipment, 70–74
Extrusion equipment, high viscosity, availability, 77–81

F

Fasteners, peel stoppers, 55
Film adhesive, performance, 38
Film adhesives, 36–40, 61
  epoxies, 36–37
  nitriles, 37
  olefin copolymers, 37–39
  phenolics, 39
  polyamides, 39
  polyolefins, 39–40
  polyvinyl butyral, 40
  polyvinyl chloride (PVC), 40
Film thickness, 38
Flame cleaning, 102
Fokker bond tester, 128
Food, Drug and Cosmetic Act, 153
Food packaging industry, 152–155

G

Glossary, 161–168
Grit blasting, 91
Guidelines for equipment selection, 64–65

H

Hand applicators, 85–86
Hardener systems for epoxy adhesives, typical (*table*), 5
Hardwoods, 47
Health hazards, 95
Heat activated (one-part system), 14–20
  benzimidazoles, 15–16
  epoxies, 16–17
  nylon, 17–18
  phenolics, 18
  polyvinyl acetals, 18–19
  rubber compounds, 19
  urethanes, 19
High bonding temperature, 38

High speed packaging system, 154
High viscosity multiple component equipment, availability, 78–81
Honeycomb materials, 54
Honeycomb sandwich construction, 36, 54–55, 157–159
Hot melt adhesive applicators, 84
Hot melt adhesives bonding, 121–122
Hot melt adhesives, comparison of, (*table*), 29
Hot melt adhesives, food packaging, 152–155
Hot melt equipment cost, 122
Hot melt types (cooling), 27–35, 61
  asphalt, 34–35
  ceramics, 35
  olefin copolymers, 30–31
  phenoxies, 31
  polyamides, 31–32
  polyesters, 32–33
  polyethylene and polypropylene, 33
  polyvinyl acetate (PVA), 33
  polyvinyl butyral, 33–34
  polyvinyl ethers, 34
Hot-setting adhesives, 119
Humidity tests, environmental, 138–140
Hydraulic cleaning, 102
Hydroblasting (*see* Hydraulic cleaning)

I

Impact test, 138
Inert gas crosslinking, 41
Inspection requirements, 123–124
Internal mix nozzles, 118
Ionomers, 31, 38

J

Joint, requirements of, 55
Joint
  service environment, 56
  strength of, 55
  weight, 56
Joint design, 49
Joints
  angle, 50
  corner, 49, 52–53
  corrugated backing, 52
  hat section, 52
  landed scarf tongue and groove, 51
  lap, 53
    double butt, 53
    double scarf, 53–54
    end, 52–53
    joggle, 53
    simple, 53
    tapered single, 53
  recessed butt, 51
  scarf tongue and groove, 51
  straight butt, 51
  T sections, 51–52

K

Kayara gum, 26
Kits, 75
Kits, sample, 75

L

Lamb waves, 128
Laminates, metal, 38, 39
Lead peroxide, 156
Leak tests, 128
"Liquid lock washer," 41

M

Manpower, 67–68
Manufacturing area size, 68–76
Materials, honeycomb, 54
MDI (*see* Diphenylmethane diisocyanate)
Melamine-paper, 48
Metal laminates, 38
Metals, surface preparation, 89–90
Metering unit, high viscosity extrusion, 71–72

N

New technology, 41
N-cyclohexyl-p-toluene-sulfonamide, 36
Nitriles
  epoxy, 37
  phenolic, 37
  vinyl, 37
Nondestructive carton opening, 153–154
Nondestructive tests, 127–128
  appearance, 127
  proof and leak, 128
  radiography, 128
  tap, 127
  ultrasonic, 127–128
Nonmetallic materials, surface preparation, 90
Novolac, 7
Nozzles
  dispensing, 78
  disposable, 75–76
  palm grip, 77–78
  spray gun, 118
    external mix, 118
    internal mix, 118
Nylons, 17, 39

O

O/p-toluenesulfonamide, 36
Olefin copolymers, 30–31, 37

One-component systems, 41
Overpackaged product, 154–155

P

Packaging, hot melt, 152–155
"Peel" loads, 55
"Peel stopper" fasteners, 55
Peel strength, 36, 38
Peel tests, 134–136
Peel values of ethylene-based adhesives (*table*), 30
Performance of film adhesives, 38
Phenolic alloys, 8
Phenolic-canvas, 48
Phenolics, 4, 18, 22, 39
Phenoxies, 31
Pickling, 102
Plant facilities, 68–87
Plasticizers, 35–36
Plywood resins, 39
Polyamides, 31, 39
Polyester-fiberglass, 48
Polyesters, 9, 32
  saturated, 8
  unsaturated, 9
Polyethylene, 28, 33
Polyolefins, 39
Polypropylene, 33
Polysulfide-epoxy adhesive properties (*table*), 10
Polysulfide liquid polymer-based sealant, 147–148
Polysulfides, 9
Polytetrafluoroethylene (TFE) cleaning, 99–100
Polyurethanes, 10
Polyvinyl acetals, 18
Polyvinyl acetate (PVA), 33
Polyvinyl butyral, 33–34, 40
Polyvinyl chloride (PVC), 40
Polyvinyl ethers, 34
Polyvinyl methyl ethers (PVME), 34
Preparation of wood as an adherend, 102–103
Pressure sensitive types, 40–41, 61
Prefit check, 124
Pretreatment techniques, 41
Primers, adhesive, 125
Production rate, 67, 144, 155
Profile height (*table*), 93
Proof tests, 128
Properties of a cured one-component epoxy polymer (*table*), 17
PVA (*see* Polyvinyl acetate)
PVC (*see* Polyvinyl chloride)
PVME (*see* Polyvinyl methyl ethers)

Q

Quality control, 123

R

Radiographic inspection, 128
Receiving inspection requirements, 123-124
Reliability, 123
Resin-glass laminates (*table*), 13
Resole, 7
Resorcinol-formaldehyde, 8
Rheology, 45
Roll coaters
  air blade, 114
  bench, 108-109
  dip roll, 113
  engraved, 114
  floor mounted, 109–110
  flow or curtain, 114–115
  "kiss" type, 110–111
  pressure or squeeze, 111–112
  reverse roll, 112–113
  two-roll reverse roll, 112–113
Rubber
  natural, 20, 26
  reclaimed, 20–21, 26
  synthetic, 21, 26
Rubber compounds, 19

S

Safety considerations, 63
Safety requirements, 152
  protective clothing, 68
  skin care, 68
  solvent cans, 68
  ventilating equipment, 68
Sandblasting method, 92–93
Sandwich structure tests, 141
Setting characteristics, 61–62
Shear and lap shear tests, 132–134
Shear strength, benzimidazole-to-metal (*table*), 16
Silicones, 12, 14
Size, manufacturing area, 68–76
Softwoods, 47
Solvent-based systems, 20–25
Specialty glue applicators, 85
Specifications (*see* American Society for Testing and materials (ASTM))
Spray application equipment, 115–118
Spray guns, 117–118
  special purpose, 118
Spray patterns and nozzles, 118
Strength values for a typical epoxy adhesive (*table*), 1

Structural adhesives, 49
Structural adhesives—aluminum bonding (*table*), 60
Structural bonding, 143–147
Substrates, dissimilar
  nonporous materials, 49
  porous materials, 48–49
Substrates, similar
  nonporous materials, 48
  porous materials, 47–48
Surface preparation, 124–125
  for acetals, 100–101
  for aluminum, 96–97
  for copper alloys, 98
  for elastomers, 98–99
  for fluorocarbons, 99
  for metals, 89–90
  for mild steel, 98
  for nonmetallic materials, 90
  for polyethylene, 101
  for stainless steel, 97–98

T

Temperature tests, environmental, 138–140
Tensile tests, 129–131
TFE (*see* Polytetrafluoroethylene)
Test exposures, standard (*table*), 140
Tests, bond
  destructive tests, 128–141
    cleavage, 136–137
    environmental corrosion, 141
    environmental temperature and humidity, 138–140
    impact, 138
    peel, 134–136
    sandwich structures, 141
    shear and lap shear, 132–134
    tensile, 129–131
  nondestructive tests, 127–128
    appearance, 127
    proof and leak, 128
    radiography, 128
    tap, 127
    ultrasonic, 127–128
Thermal coefficient, 45
Triphenylmethane triisocyanate (TTI), 23
TTI (*see* Triphenylmethane triisocyanate)

U

Ultrasonic tests, 127–128
Urethanes, 14, 19, 22
U. S. Forest Products Laboratory, 135
U. S. Naval Ordnance Laboratory, 130

V

Vinyl resins, 23, 26

W

Waste disposal, 86–87
Water-based systems, 25–27
Water solubility, 34
Wet blasting, 93–94
Wettability, adhesive, 45, 90